# Ground Improvement Techniques

# Ground Improvement Techniques

**Bujang B. K. Huat**

*Universiti Putra Malaysia, Selangor, Malaysia*

**Arun Prasad**

*Indian Institute of Technology (BHU), Varanasi, India*

**Sina Kazemian**

*Payame Noor University (PNU), Tehran, Iran*

**Vivi Anggraini**

*Monash University Malaysia, Bandar Sunway, Malaysia*

CRC Press
Taylor & Francis Group
Boca Raton London New York Leiden

CRC Press is an imprint of the
Taylor & Francis Group, an **informa** business

A BALKEMA BOOK

*CRC Press/Balkema is an imprint of the Taylor & Francis Group, an informa business*

First issued in paperback 2021

© 2020 Taylor & Francis Group, London, UK

Typeset by Apex CoVantage, LLC

*Library of Congress Cataloging-in-Publication Data*

Applied for

Published by:   CRC Press/Balkema
                Schipholweg 107c, 2316 XC Leiden, The Netherlands
                e-mail: Pub.NL@taylorandfrancis.com
                www.crcpress.com – www.taylorandfrancis.com

ISBN-13: 978-1-138-54103-0 (hbk)
ISBN-13: 978-1-03-208542-5 (pbk)
ISBN-13: 978-0-429-50765-6 (ebk)

DOI: https://doi.org/10.1201/9780429507656

# Contents

# Chapter 1

# Introduction

*An approximate solution to the right problem is more desirable than a precise solution to a wrong problem.*

—*Anonymous (2001)*

## 1.1    Introduction

Different regions may have different soils with geotechnical problems altogether different from those faced in other regions. For example, in Southeast Asia, the common geotechnical problems are those associated with construction with soft clays, organic soils and peat, while in the arid region of the Middle East, problems are generally associated with the desert (dry) soils. In the United States, there are problems associated with organic soils, expansive and collapsing soils and shale. Laterite and lateritic soils are especially problematic in Mexico. Similarly, in the European Union, geotechnical problems are associated with loess (France) and organic soil (Germany). Figure 1.1 shows a map of the global soil regions.

In Southeast Asia, central Africa, and North and South America, the major soil type is utisols. Utisols, commonly known as "red clay" soils, are soils that have formed in humid areas and are intensely weathered. In India, the major soils types are utisols and vertisols. Vertisols are soils in which there is a high content of expansive clay, known as montmorillonite, which forms deep cracks in drier seasons or years. Vertisols typically form from highly basic rocks, such as basalt. In the case of the Middle East, parts of the United States and eastern Australia, the main soil type is aridisols. Aridisols (or desert soils) are formed in an arid or semi-arid climate. Aridisols dominate the deserts and xeric shrublands which occupy about one-third of the Earth's land surface. In central Africa and the northern part of South America, oxisols are the common soil type. The main processes of soil formation of oxisols are weathering and humification. Oxisols are always a red or yellowish color due to the high concentration of iron, aluminum oxides and hydroxides. They also contain quartz and kaolin, plus small amounts of other clay minerals and organic matter.

In central Asia, part of the United States, Argentina and Brazil, the common soil type is mollisols. Mollisols form in semi-arid to semi-humid areas, typically under a grassland cover. They are most commonly found in the mid-latitudes: mostly east of the Rocky Mountains in North America; in Argentina (the Pampas) and Brazil in South America; and in Mongolia and the Russian steppes of Asia. Their parent material is typically base-rich and calcareous and includes limestone, loess, or windblown sand. Meanwhile, in the cold north (Alaska and Siberia), gelisols are found. They are soils of very cold climates which are defined as containing permafrost within 2 m of the soil surface.

# Global Soil Regions

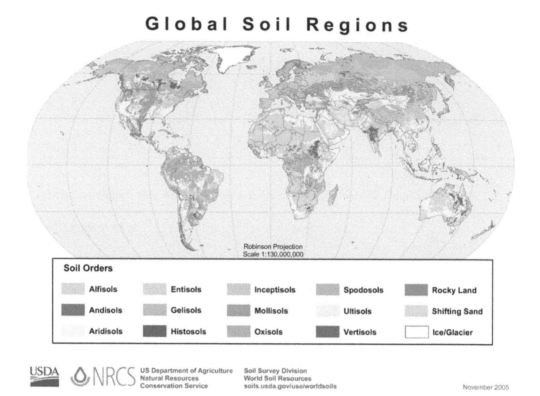

*Figure 1.1* Global soil regions

Source: https://en.m.wikipedia.org/wiki/Soil_science#/media/File%3AGlobal_soil_regions.jpg

A brief account of geotechnical problems for some regions is presented next for example. In the case of Southeast Asia, there are two distinctive different geographic regions. The first is mainland Southeast Asia, also known as Indochina, on the Indochinese peninsula; it comprises Cambodia, Laos, Myanmar (Burma), Thailand, Vietnam and West Malaysia (Peninsular Malaysia). The second region is the Malay Archipelago, or Maritime Southeast Asia, which comprises Brunei (on the island of Borneo), East Malaysia (with the Malayan states of Sabah and Sarawak on the northern part of Borneo), all the islands of Indonesia, the Philippines, Singapore and Timor-Leste (East Timor) (www.nationsonline.org). A map of Southeast Asia is shown in Figure 1.2.

The soil engineering problems of Southeast Asia are normally associated with soft ground, namely soft clays (alluvial and marine), silt and shale formation as well as organic soils and peat. Alluvial soils are also found in many coastal regions of the world.

The coastal plain of Peninsular Malaysia, for example, is of very low relief, standing at most a few meters above sea level. The width of the coastal plain varies greatly and can be up to 20 km or more along relatively large stretches of both eastern and western coasts. Large deposits of weak marine clays are encountered throughout Peninsular Malaysia. They are found in Johor, Malacca, Klang, Penang and Alor Star. The properties of Malaysian marine clay soil vary significantly for moist and dry soils.

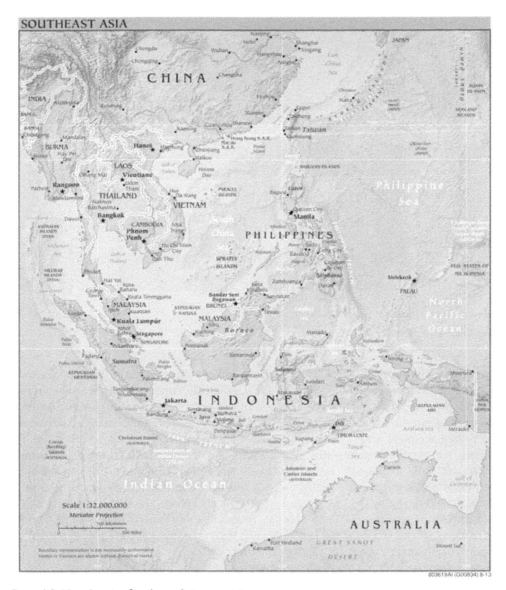

*Figure 1.2* Map showing Southeast Asian countries

Source: https://en.wikipedia.org/wiki/Southeast_Asia#/media/File:Political_Southeast_Asia_Map.jpeg

Indonesia consists of the main islands of Java, Sumatra, Sulawesi, Irian Jaya and part of Borneo (Kalimantan). Expansive soils are widely distributed in central to west Java, Indonesia. Beside of its swelling potential, expansive soils present in a region of Yogyakarta are highly susceptible to the action of weathering. Under natural conditions, weathering causes soil disintegration and leads to deterioration of its physical and mechanical properties (Muntohar, 2006). Expansive soils may cause heavy distress to engineering constructions.

*Figure 1.3* Deterioration of clay shale formation
Source: Oktaviani et al. (2018)

Foundation soils which are expansive will heave and can cause lifting of foundations and other structures during periods of high moisture. Conversely, during periods of falling moisture, expansive soils can collapse, thereby resulting in building settlement. The availability of loose sand deposits along the coastal line of Indonesia is also life-threatening as Indonesia is an earthquake-prone area.

Apart from the problematic soft clay and loose sand deposits, Indonesia has also many problematic shale formations. These formations are primarily composed of expandable, smectitic minerals, pyrite and soluble minerals of calcite and siderite (Irsyam, Susila, and Himawan, 2007). During the soil investigation, when unexposed, the shale gives very high standard penetrometer test (SPT) blow counts (normally more than 40–50 blows/ft). However, once excavated, exposed and in contact with water, the strength can easily deteriorate, as shown in Figure 1.3.

Peat is abundantly found along the eastern coast of Sumatra and western coast of Kalimantan. Peat is estimated to cover 13% of the total land area of Indonesia, one-third of which is found in Kalimantan. The peat of Kalimantan is characterised by a low nutrient status and a low pH. Peat is also found in many countries throughout the world. In the United States, peat is found in 42 states, with a total acreage of 30 million hectares (ha). Canada and Russia are two countries with a large area of peat – 170 and 150 million ha, respectively. In Malaysia, some 3 million ha (about 8%) of the country's land area is covered with peat. Peat has certain characteristics that sets it apart from most mineral soils and requires special considerations for construction over them. These special characteristics include (Edil, 1994):

1.  High natural moisture content (up to 1500%)
2.  High compressibility, including significant secondary and tertiary compression

3.   Low shear strength (typically $S_u = 5$–$20$ kPa)
4.   High degree of spatial variability
5.   Potential for further decomposition as a result of changing environmental conditions.

Singapore, located in a tropical climate, is dominated by residual soils from two major geological formations: the Bukit Timah granitic formation and the Jurong sedimentary formation. These residual soils constitute two-thirds of Singapore's land area (Public Works Department, 1976). About 25% of the land area of Singapore is underlain by recent deposits of marine and alluvial origin. The most common of their recent deposits is Singapore marine clay, which is up to 40 m thick. This soft clay is a Quaternary deposit that lies within submarine valleys cut in an old alluvial formation and is locally known as the Kallang formation. Problems often arise in construction in Singapore involving this soft marine clay owing to its low shear strength and high compressibility. The Singapore peaty soils are usually found together with the soft soil deposits of the Kallang formation. The younger peaty soils are usually found at shallow depth of 3–5 m above the upper marine clay, while the older peaty soils are found at a depth of 10 m sandwiched between the upper and lower marine clays (Tan, 1983).

In poor and weak subsoils, the design of a conventional shallow foundation for structures may present problems with respect to both sizing of the foundation as well as control of foundation settlements. Traditionally pile foundations have been employed, but they prove to be too expensive. A more viable alternative to pile foundation in certain situations is to improve the subsoil itself to such an extent that the subsoil would develop an adequate bearing capacity; foundations constructed after subsoil improvement would have resultant settlements within acceptable limits.

Constructions on soft ground, such as building embankments for road, railway or landfill for housing project and building foundations, may bring problems associated with instability during construction and long-term and persistent settlement thereafter, as illustrated in Figures 1.4 and 1.5.

In India, alluvial soils are found in few states in the northern area, as shown in Figure 1.6. These soils include the deltaic alluvium, calcareous alluvial soils, coastal alluvium and coastal sands. Alluvial soils occur chiefly in the Indo-Gangetic plain covering the states of Punjab and Haryana in the northwest, Uttar Pradesh and Bihar in the north, and West Bengal and parts of Meghalaya and Odisha in the northeast. Alluvial soils are generally made up of fine silt brought down by rivers from the mountains. Structures constructed on cohesive or alluvial soils may be subject to large settlement due to the compressible nature of the foundation soil. Marine soils are found in a narrow belt near the southwest coast of India. These soils have low shear strength and high compressibility. They contain a large amount of organic matter and are not suitable for construction of megastructures. Another major soil type, which covers about 20% of the total area of the country is the black soils (black cotton soils, or vertisols). This soil, which is also known as *regur* soil, is distributed over the Deccan lava tract which extends over parts of Maharashtra, Madhya Pradesh, Gujarat, Andhra Pradesh, Karnataka and Tamil Nadu. The black color is due to the presence of titaniferous magnetite. These soils cover an area of about 5,460,000 km². Black soils are predominantly clay with patches of clay loams, loams and sand loams. Black soil varies in depth from shallow to deep. It is derived from two types of rocks, the Deccan and Rajmahal trap and ferruginous gneisses and schists occurring in Tamil Nadu state under semi-arid conditions. The former attains considerable depth whereas the latter are generally shallow. When these soils

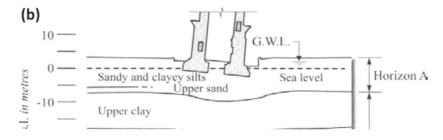

*Figure 1.4* Bearing capacity failure and instability of soft ground

Source: www.pwri.go.jp

**(a)**

**(b)**

*Figure 1.5* Long-term settlement problem: (a) Leaning Tower of Pisa (Italy); (b) soil profile

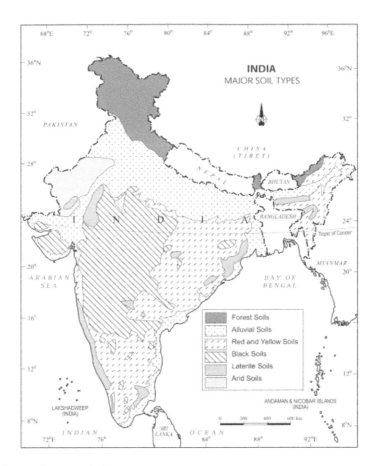

*Figure 1.6* Major soil types in India

Source: http://ncert.nic.in/ncerts/l/kegy106.pdf

become wet, they will expand. The resulting expanding pressure can cause uplift against concrete slabs and foundation footings, causing a wide variety of damage to buildings and surrounding areas. The damages may include cracking and heaving of concrete, including garage slabs, sidewalks, foundations, basement walls and basement floor slabs. It can also lead to water leaking into basements, broken pipes and water lines, cracks in interior walls and sticking doors and windows. Black cotton soil has very low shear strength, low California bearing ratio (CBR) (soaked CBR = 1–2), liquid limit of 40%–100% and low permeability in its dry state. This results in usually very large settlement. The black cotton soil is very hard when dry, but loses its strength completely when wet. Apart from the Deccan Plateau of India, black cotton soils are also dominant in eastern Australia (especially inland Queensland and New South Wales); parts of southern Sudan, Ethiopia, Kenya, Chad and northeast Nigeria in Africa; the lower Paraná River in South America; southern Texas and adjacent Mexico in North America; Thrace and New Caledonia in Europe; and parts of eastern China.

Red soils (utisols) are another major soil type of India. They cover a large area in the south and northeast of the peninsula. Such soils occur in Tamil Nadu, eastern Andhra Pradesh,

Karnataka, Goa, southeastern Maharashtra, Madhya Pradesh and parts of Jharkhand, Odisha, Uttar Pradesh and West Bengal. Red soils cover an area of about 90,000 km². The ancient crystalline and metamorphic rocks on meteoric weathering have given rise to red soils. The reddish color of utisols is due to presence of iron in the crystalline and metamorphic rocks. Red soils are mainly sandy to loamy in texture, with gravels on upper slopes, then sandy soils, deeper loamy soils on slopes and loams or rather clay in the valley bottoms. These soils are found in areas with low rainfall and they are not capable of retaining moisture. Red soil possesses lower strength compared to other soils due to its porous and friable structure.

The Middle East, part of Southern Africa and central Australia can be generally classified as desert regions. The subsoil in desert regions (aridisols) is a product of temperature changes, wind and rainfall. A climate of high temperatures increases evaporation and reduces moisture in the soil, resulting in development of salt-bearing soil. A map of the world's major deserts is shown in Figure 1.7.

The geotechnical problems associated with deserts can be broadly put in the following categories:

* Expansive soil due to presence of some minerals with expansive potential
* Collapsible soils
* Presence of inland and coastal salt-bearing soil (Sabkha soil)
* Compressible and shrinking loose Aeolian deposits
* Shifting sand dunes
* Gypseous soils due to presence of hydrated gypsum in poorly drained areas
* Presence of highly weathered layers and variability of foundation material
* Presence of cavities in limestone formations.

Expansive soils are soils that expand when water is added and shrink when they dry out. This continuous change in soil volume causes homes built on this soil to move unevenly and crack (Figure 1.8).

*Figure 1.7* Largest deserts in the world

Source: www.quickgs.com/major-deserts-of-world/

(a) Typical cracking of expansive soil when dry

(b) Case in Saida city – Algeria

*Figure 1.8* Expansive soil: (a) typical picture of expansive soil, and (b) a cracked building

Collapsible soils are found in many countries in the Middle East (e.g., Saudi Arabia) and South and North Africa (e.g., Egypt, Algeria). These soils are also found in many other countries including the United States, Russia, China, South America (e.g., Brazil), and many countries in eastern Europe. In general, collapsible soils are located in arid and semi-arid regions around the world. This special type of soil is characterised by abrupt reduction in strength and excessive and sudden settlement when it becomes wet, leading to failure of the structure (Figure 1.9). Construction on such soil is one of the prominent problems in geotechnical engineering.

Sabkha (or Sabkha soils) are soils with a high concentration of salts. These soils originate due to capillary suction and intense evaporation in the coastal and inland flat plains. Sometimes the salinity of the pore fluid reaches as high or higher than that of seawater. The high salt content has a great impact on the strength properties of soils and also on structures in contact with these soils. The collapse potential of saline soils is principally related to the dissolution of salts. These soils have very low strength, low bearing capacity and high compressibility. The expected settlement is always above the recommended tolerable limits. In addition, the salts are highly corrosive. Figure 1.10 shows a picture of Sabkha soils.

Compressible and shrinking, loose Aeolian deposits are desert deposits which exhibit high compressibility. In a dry state, the deposits are relatively incompressible with appreciable shear strength, but when exposed to excessive water the deposits lose strength and undergo excessive compression, often leading to collapse of structures. These sand tracts are not only a major hazard to highways and other communication systems but also have other problematic properties typical of Aeolian deposits. Global distribution of Aeolian deposits is shown

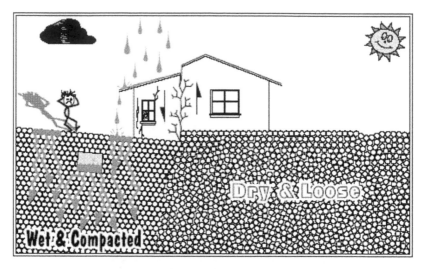

*Figure 1.9* A sketch of foundation failure in collapsible soil

Source: www.aegweb.org

(a) Coastal Sabkha in Saudi Arabia (www.jubail-wildlife-sanctuary.info)

(b) Chott Merouane, Algeria (www.yannarthusbertrand2.org)

*Figure 1.10* Pictures showing Sabkha soils

in Figure 1.11, and typical Aeolian deposits in the United Arab Emirates (UAE) and Algeria are shown in Figures 1.12 and 1.13, respectively.

Sand dunes move forced by wind through different mechanisms. They can move through a mechanism known as saltation, where the particles of sand are removed from the surface

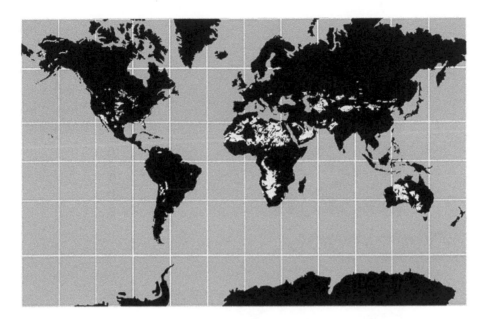

*Figure 1.11* Global distribution of major deposits of Aeolian-derived sediments

Source: www.physicalgeography.net

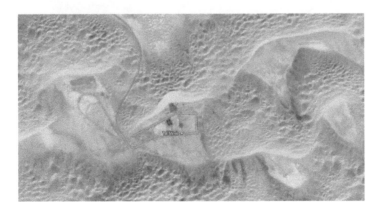

*Figure 1.12* Compound crescentic dunes in Liwa, UAE

Source: Christopher et al. (2016)

and are carried by the wind before landing back on the surface. When these particles land, they can scatter other particles and cause them to move as well. Another mechanism is present on the steep slopes of the dunes, where the sand is falling down: this is the "sand avalanche". Therefore, if we are on the sloping windward side, we can see the sand grains that jump a few centimeters above the surface of the dune. At the dune's crest, the airborne sand grains fall down the steep slope as small avalanches. With strong winds, the sand particles

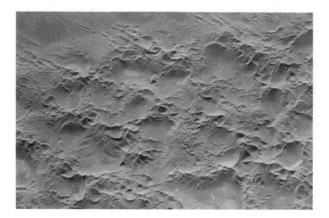

*Figure 1.13* The Issaouane Erg in Algeria

Source: www.earthobservatory.nasa.gov

*Figure 1.14* A typical shifting sand dune

Source: www.jamso.me

move in a sheet flow. This is an overland motion of the sand having the form of a continuous layer over the soil. The mass transferred by this flow is extremely large. Then, during a strong dust storm, the dunes may move more than several meters (Figure 1.14). A map showing sand dunes in the Arabian peninsula is shown in Figure 1.15. Sand dunes affect many infrastructures and cause delays in construction.

Gypsum present in gypseous soil causes apparent cementation when the soil is dry. Nevertheless, the addition of water results in dissolution and often leads to serious collapse. In Iraq, for example, the percentage of gypsum content varies from 10% to 70%, with about a third of the country's land area covered with gypseous soil. In areas when the annual rainfall exceeds 350 mm, the gypsum content typically ranges between 3% and 10%. In arid regions

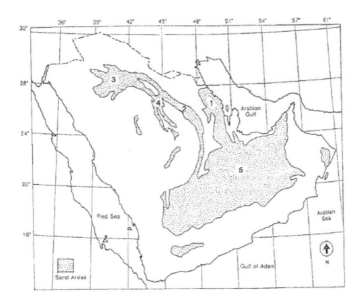

*Figure 1.15* Sand terrains in the Arabian Peninsula

Source: Al-Refeai and Al-Ghamdy (1994)

(with less than 250 mm of rainfall annually) the gypsum percentage typically exceeds 50%. The problems encountered in gypseous soils can be summarised as follows:

- Great loss in strength upon wetting
- Sudden increase in compressibility upon wetting
- Continuation of deformation and collapse upon leaching due to water movement
- The existence of cracks due to seasonal changes
- The existence of sinkholes in the soil due to local dissolution of gypsum.

Sinkholes are common where the rock below the land surface is limestone, carbonate rock, salt beds or rocks that can naturally be dissolved by groundwater circulating through them. As the rock dissolves, spaces and caverns develop underground. Sinkholes are dramatic because the land usually stays intact for a while until the underground spaces just get too big. If there is not enough support for the land above the spaces, then a sudden collapse of the land surface can occur. These collapses can be small or very large, and can even occur beneath a house or road. Figure 1.16 shows the problems of sinkholes formed in limestone in Algeria.

The Islamic Republic of Iran is located in Western Asia, bordering the Gulf of Oman, the Persian Gulf, and the Caspian Sea, and covering a land area of 1,636 million square kilometres. It borders on: Afghanistan, Armenia, Azerbaijan, Iraq, Pakistan, Turkey and Turkmenistan. Much of Iran sits on the Iranian plateau and is at high altitude (www.fao.org). Figure 1.17 shows a map of Iran with its deserts.

The major soil types of the European Union are shown in Figure 1.18. The European Union (EU) is a political and economic union of 28 member states that are located primarily in Europe.

A map showing the major soil types of the United States is presented in Figure 1.19.

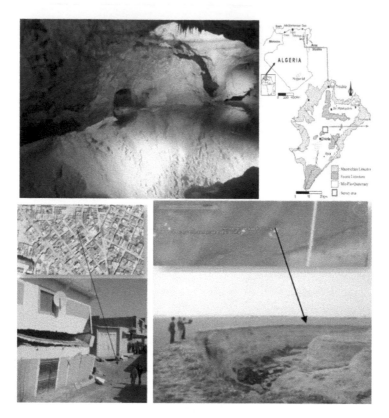

*Figure 1.16* Pictures showing sinkholes in limestone in Algeria
Source: Fehdi et al. (2014); Azizi et al. (2014)

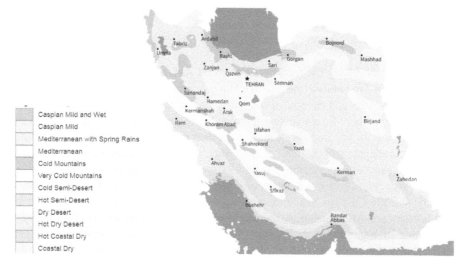

*Figure 1.17* A map of Iran with its deserts
Source: www.wikipedia.org

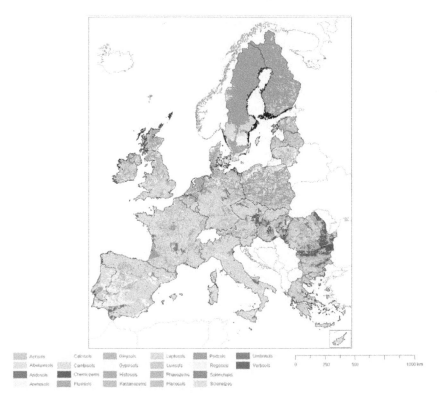

*Figure 1.18* Major soil types of Europe

Source: http://citeseerx.ist.psu.edu

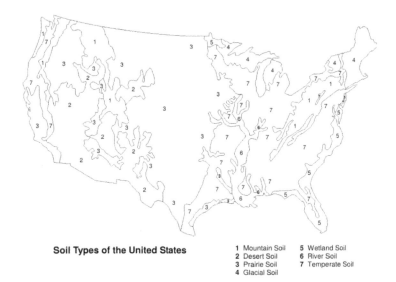

**Soil Types of the United States**

| 1 | Mountain Soil | 5 | Wetland Soil |
|---|---------------|---|--------------|
| 2 | Desert Soil | 6 | River Soil |
| 3 | Prairie Soil | 7 | Temperate Soil |
| 4 | Glacial Soil | | |

*Figure 1.19* Soil types of the United States

Source: www.thudscave.com

## 1.2   Ground improvement techniques – a brief introduction

With the development of science and technology, we now have a number of alternative construction methods. It is also perhaps fair to say that geotechnical engineering is more of an art than an exact science. This is because we have to deal with natural materials like soil and rock, and we have no control in their formation; they are materials whose properties are difficult to model with simple mathematics. Therefore, it must be realised that a particular method (say, soil improvement) developed for other countries may not necessarily work for local conditions. Because of this, it is prudent to thoroughly test the method first, to study its effectiveness before accepting it for local practice. The techniques for ground improvement have developed rapidly in recent years and have found large-scale application in industrial projects.

In general, ground improvement is aimed at:

1.  Methods to improve the engineering properties of the treated soil mass
2.  Implemented by modifying the ground's character – with or without the addition of foreign material
3.  Properties modified are shear strength, stiffness and permeability
4.  The major function of ground improvement is to:

    a.  Increase the bearing capacity
    b.  Control deformations and accelerate consolidation
    c.  Provide lateral stability
    d.  Form seepage cut-off and environmental control
    e.  Increase resistance to liquefaction.

What follows are some of the ground improvement methods that are commonly used, depending on soil types and regions.

For the Southeast Asian region (and other similar regions), for example, geotechnical problems are generally associated with soft ground, namely soft clays (alluvial and marine) and silt, as well as organic soils and peat. For the soft ground, the commonly used ground improvement methods include preloading, drainage (including vertical drains), vibro-compaction, dynamic compaction, reinforcement (sand and stone columns), grouting and treatment with chemicals such as lime, electro-osmosis, and several others. In the case of organic soils and peat, this may include partial or total removal for shallow peat, preloading, vertical drains, geotextile separator and bamboo or timber fascine/mattresses. In the case of expansive soils, one method is to protect the soil mass from excessive wetting and drying (moisture control).

For the black cotton soil of India, the engineering properties of the soil can significantly be improved with treatment with lime or cement. Cement or hydrated lime in the range of 3–5% can bring remarkable improvement in the engineering characteristics of the black cotton soil. Cement- and lime soil stabilisation technology has been found useful, cost-effective and suited to manual methods of construction. This technology is 20%–30% cheaper than conventional water bound macadam (WBM) construction. The cement or lime treatment is being utilised for the following purposes:

1.  To provide a pavement foundation of marginally weaker strength than that of concrete pavement, but much improved strength compared to natural black cotton soil.
2.  To consolidate subgrades and base courses for concrete pavement in order to make them resistant to volume changes and displacement or erosion in the presence of moisture even under the rocking action of curled slabs, if any.
3.  To overcome the susceptibility of foundations to volume change and to increase their shearing resistance and bearing capacity.

For the Middle East and arid regions, the geotechnical problems are generally related to desert soils such as loess, shifting sand dunes, expansive and collapsible soils, Sabkha and gypseous soils.

In general, the following ground improvement methods are used:

**Expansive soils**

1.  Removal: The soils engineer may recommend that the uppermost several feet of expansive soil be removed and new non-expansive material be imported and compacted to create a stable layer of soil at the building footprint.
2.  Partial replacement: Depending upon the severity of expansion potential, non-expansive soils may be mixed with expansive soil to lower the expansion potential to an acceptable level.

**Sabkha**

1.  Using drainage (open channel) systems with saline inflows to manage the discharge of the saline.
2.  Protecting the building foundations against the saline attack by using sulfate-resistant cement or by using insulators like bitumen.
3.  Using chemical agents to reduce soil salinity.

**Gypseous soils**

Gypseous soils can be improved by decreasing the effect of water on the soils to ensure the safety and stability of the engineering structures. This treatment can be achieved chemically or physically.

1.  The physical treatment means that the soil properties are improved using mechanical methods.
2.  Soil replacement (the preferred solution in the places where the replacement soils are readily available).
3.  Compaction (in which the soil must be wetted at first and then compacted).
4.  Soil reinforcement (using geomembrane or geogrid).
5.  Dynamic compaction.
6.  Pre-wetting (the soil must be wetted to remove the salt from the soil structure).
7.  Chemical treatment with additives such as cement, lime or bitumen.

Construction and maintenance of structures in the arctic and subarctic regions underlain by permafrost are characterised by a wide range of problems in addition to those experienced elsewhere. Engineers, designers, and construction and maintenance personnel are continuously plagued by extremely severe frost heaving, subsidence caused by thawing permafrost, soil creep or solifluction, landslides and icing. These processes not only present construction and maintenance problems but in the more severe cases may be a hazard to the users of the structures. There are two basic methods of constructing on permafrost:

1.   The active method
2.   The passive method.

The active method is used in areas where permafrost is thin and generally discontinuous or where it contains a relatively small amount of ice. The object of this method is to thaw the permafrost, and if the thawed material has a satisfactory bearing strength, then construct in a normal manner. Sometimes the structure itself can be used to thaw the permafrost and the surface brought back to grade at intervals while the thawing process takes place. Generally, the thawing of the permafrost is accomplished simply by clearing the vegetation, which normally insulates the permafrost from the heat in the air and from solar radiation. Heat from steam and warm water piped into the ground has also been used to thaw permafrost. Naturally, the active method of construction has limited application because of the great thickness of permafrost throughout most of the region and because of the time required for thawing the ground.

The passive method, which has broad application throughout most of the permafrost region, is used in areas such as the North Slope of Alaska, where it is impractical to thaw the permafrost. The object of this method is to minimise disturbance of the permafrost and of the thermal regime. The thermal regime in an area in a natural undisturbed state is normally in quasi-equilibrium with all the environmental factors, but in many areas this state of equilibrium is very sensitive. The simple passage of a tracked vehicle that destroys the vegetation mat is enough to upset the delicate balance and cause the top of the permafrost layer to thaw. This thawing can cause differential settlement of the surface of the ground, drainage problems, and severe frost action. Once the equilibrium is upset, the whole process can feed on itself and be practically impossible to reverse. However, if a structure is founded on permafrost that remains frozen, the frozen ground provides rocklike bearing strength. In special cases it may be practical to keep the permafrost intact by using a refrigeration system. Engineering problems in permafrost nearly always are associated with the active layer (the layer that freezes and thaws annually) and the active layer permafrost interface (permafrost table). Changes in the surface environment, either natural or man-made, produce thermal changes in this zone that can have serious effects upon engineering works and structures.

There are several standards and guidelines (e.g., IS 13094–1992) available that provide a summary of the various ground improvement methods that can be considered, as shown in Table 1.1:

Table 1.1 Summary of soil improvement methods

| Type of Soil | Method | Principle | Most Suitable Soil Conditions/Types | Maximum Effective Treatment Depth | Special Materials Required | Special Equipment Required | Properties of Treated Material | Special Advantages and Limitations | Relative Cost |
|---|---|---|---|---|---|---|---|---|---|
| **In Situ Deep Compaction of Cohesionless Soils** | Blasting | Shock waves and vibrations cause liquefaction and displacement with settlement to higher density | Saturated, clean sands; partly saturated sands and silts (collapsible loess) after flooding | >30 m | Explosives, backfill to plug drill holes, hole casings | Jetting or drilling machine | Can obtain relative densities to 70–80, may get variable density time-dependent strength gain | Rapid, inexpensive, can treat any size areas; variable properties, no improvement near surface, dangerous | Low |
| | Vibratory Probe | Densification by vibration; liquefaction-induced settlement under overburden | Saturated or dry clean sand | 20 m (ineffective above 3–4 m depth) | None | Vibratory pile drived and 750 mm diameter open steel pipe | Can obtain relative densities of up to 80. Ineffective in some sands | Rapid, simple, good underwater, soft underlayers may damp vibrations, difficult to penetrate stiff overlays, not good in partly saturated soils | Moderate |
| | Vibro-compaction | Densification by vibration and compaction of backfill material | Cohesionless soils with less than 20% fines | 30 m | Granular backfill, water supply | Vibrofloat, crane, pumps | Can obtain high relative densities, good uniformity | Useful in saturated and partly soils, uniformity | Moderate |
| | Compaction Piles | Densification by displacement of pile volume and by vibration during driving | Loose sandy soils; partly saturated clayey soils, loess | >20 m | Pile material (often sand or soil plus cement mixture) | Pile driver, special sand, pile equipment | Can obtain high densities, good uniformity | Useful in soils with fines, uniform compaction, easy to check results, slow, limited improvement in upper 1–2 m | Moderate to high |
| | Heavy Tamping (Dynamic Consolidation) | Repeated application of high intensity impacts at surface | Cohesionless soils, waste fills, partly saturated soils | 30 m | None | Tampers of up to 200 tons, high capacity crane | Can obtain good improvement and reasonable uniformity | Simple, rapid, suitable for some soils with fines; usable above and below water, requires control, must be away from existing structures | Low |

(Continued)

Table 1.1 (Continued)

| Type of Soil | Method | Principle | Most Suitable Soil Conditions/ Types | Maximum Effective Treatment Depth | Special Materials Required | Special Equipment Required | Properties of Treated Material | Special Advantages and Limitations | Relative Cost |
|---|---|---|---|---|---|---|---|---|---|
| **Injection and Grouting** | Particulate Grouting | Penetration grouting – fill soil pores with soil, cement, and/or clay | Medium to coarse sand and gravel | Unlimited | Grout, water | Mixers, tanks, pumps, hoses | Impervious, high strength with cement grout, eliminate liquefaction danger | Low-cost grouts, high strength, limited to coarse-grained soils, hard to evaluate | Lowest of the grout systems |
| | Chemical Grouting | Solutions of two or more chemicals react in soil pores to form a gel or a solid precipitate | Medium silts and coarser | Unlimited | Grout, water | Mixers, tanks, pumps, hoses | Impervious, low to high strength with cement grout, eliminate liquefaction danger | Low viscosity controllable gel time, good water shut-off, high cost, hard to evaluate | High to very high |
| | Pressure-Injected Lime | Lime slurry injected to shallow depths under high pressure | Expansive clays | Unlimited, but 2–3 m usual | Lime, water surfactant | Slurry tanks, agitators, pumps, hoses | Lime encapsulated zones formed by channels resulting from cracks, hydraulic fracture | Only effective in narrow range of soil conditions | Competitive with other solutions to expansive soil problems |
| | Displacement Grout | Highly viscous grout acts as radial hydraulic jack when pumped in under high pressure | Soft, fine-grained soils, foundation soils with large voids or cavities | Unlimited, but a few meters usual | Soil, cement, water | Batching equipment, high-pressure pumps, hoses | Grout bulbs within compressed soil matrix | Good for corrections of differential settlements, filling large voids, careful control required | Low material, high injection |
| | Electro-kinetic Injection | Stabilising chemicals moved into soil by electro-osmosis or colloids into pores by electrophoresis | Saturated silts, silty, clayey (clean sands in case of colloid injection) | Unknown | Chemical stabilizer colloidal void fillers | DC power supply, anodes, cathodes | Increased strength, reduced compressibility, reduced liquefaction potential | Existing soil and structures not subjected to high pressures, no good in soils with high conductivity | Expensive |

| | Process | Suitable soils | | Materials | Special equipment | Results | Comments | Cost |
|---|---|---|---|---|---|---|---|---|
| Jet Grouting | High speed jets at depth excavate, inject, and mix stabiliser with soil to form columns or panels | Sands, silts, clays | — | Water, stabilising chemicals | Special jet nozzle, pumps, pipes and hoses | Solidified columns and walls | Useful in soils that can't be permeation grouted, precision in locating treated zones | — |
| **Precompression** | | | | | | | | |
| Preloading with/without Drain | Load is applied sufficiently in advance of construction so that compression of soft soils is completed prior to development of the site | Normally consolidated soft clays, silts, organic deposits, completed sanitary landfills | — | Earth fill or other material for loading the site, sand or gravel for drainage blanket | Earth-moving equipment, large water tanks or vacuum drainage systems sometimes used, settlement markers, piezometers | Reduced water content and void ratio, increased strength | Easy, theory well developed, uniformity, requires long time (vertical drains can be used to reduce consolidation time) | Low (moderate if vertical drains are required) |
| Surcharge Fills | Fill in excess of that required permanently is applied to achieve a given amount of settlement in a shorter time, excess fill then removed | Normally consolidated soft clays, silts, organic deposits, completed sanitary landfills | — | Earth fill or other material for loading the site, sand or gravel for drainage blanket | Earth-moving equipment, settlement markers, piezometers | Reduced water content, void ratio and compressibility, increased strength | Faster than preloading without surcharge, theory well developed, extra material handling, can use vertical drains to reduce consolidation time | Moderate |
| Electro-osmosis | DC current causes water flow from anode toward cathode where it is removed | Normally consolidated silts and silty clays | — | Anode (usually rebar or aluminum), cathodes (well points or rebar) | DC power supply, wiring, metering | Reduced water content and compressibility, increased strength, electrochemical hardening | No fill loading required, can be used in confined areas, relatively fast, non-uniform properties between electrodes, no good in highly conductive soils | High |

(Continued)

Table 1.1 (Continued)

| Type of Soil | Method | Principle | Most Suitable Soil Conditions/Types | Maximum Effective Treatment Depth | Special Materials Required | Special Equipment Required | Properties of Treated Material | Special Advantages and Limitations | Relative Cost |
|---|---|---|---|---|---|---|---|---|---|
| | Remove and Replace | Foundation soil excavated, improved by drying or admixture, and recompacted | Inorganic soils | 10 m | Admixture, stabilisers | Excavating, mixing, and compaction equipment, dewatering system | Increased strength and stiffness, reduced compressibility | Uniform, controlled foundation soil when replaced, may require large area dewatering | High |
| **Admixtures** | Structural Fills | Structural fill distributes loads to underlying soft soils | Use over soft clays or organic soils, marsh deposits | — | Sand, gravel, fly ash, bottom ash, slag, expanded aggregate, clamshell or oyster shell, incinerator ash | Mixing and compaction equipment | Soft subgrade protected by structural load-bearing fill | High strength, good load distribution to underlying soft soils | Low to high |
| | Mix-in-Place Piles and Walls | Lime, cement, or asphalt introduced through rotating auger or special in-place mixer | All soft or loose inorganic soils | >20 m | Cement, lime, asphalt, or chemical stabiliser | Drill rig, rotary cutting and mixing head, additive proportioning equipment | Solidified soil piles or walls of relatively high strength | Use native soil, reduced lateral support requirements during excavation, difficult quality control | Moderate to high |
| **Thermal** | Heating | Drying at low temperature, alteration of clays at intermediate temperatures (400–600°C), fusion at high temperature (>1000°C) | Fine-grained soils, especially partly saturated clays and silts, loess | 15 m | Fuel | Fuel tanks, burners, blowers | Reduced water content, plasticity, water sensitivity, increased strength | Can obtain irreversible improvement in properties, can introduce stabilisers with hot gases | High |

| | Method | Principle | Soils | Depth | Material | Equipment | Effect | Comments | Relative cost |
|---|---|---|---|---|---|---|---|---|---|
| Reinforcement | Freezing | Freeze soft, wet ground to increase its strength and stiffness | All soils | Several meters | Refrigerant | Refrigeration system | Increased strength and stiffness, reduced permeability | No good in flowing groundwater, temporary | High |
| | Vibro-replacement Stone and Sand Columns | Hole jetted into soft, fine-grained soil and backfilled with densely compacted gravel or sand | Soft clays and alluvial deposits | 20 m | Gravel or crushed rock backfill | Vibroflot, crane or vibrocat, water | Increased bearing capacity, reduced settlements | Faster than precompression, avoids dewatering required for remove and replace, limited bearing capacity | Moderate to high |
| | Root Piles, Soils Nailing | Inclusions used to carry tension, shear, compression | All soils | — | Reinforcing bars, cement grout | Drilling and grouting equipment | Reinforced zone behaves as a coherent mass | In situ reinforcement for soils that can't be grouted or mixed in place with admixtures | Moderate to high |
| | Strips and Membranes | Horizontal tensile strips, membranes buried in soil under embankment, gravel base courses and footings | Cohesionless soils | Can construct earth structures to heights of several tens of meters | Metal or plastic strips, geotextiles | Excavating, earth handling and compaction equipment | Self-supporting earth structures, increased bearing capacity, reduced deformations | Economical, earth structures coherent, can tolerate deformations, increased allowable bearing pressure | Low to moderate |

This book is specially written with the following aims in mind:

1.  To give a state-of-the-art review on problems during design and construction in problematic soils.
2.  To explain design methods, site investigation, construction and analysis of the various improvement methods available.

Based on these objectives, this book has been written in the following form:

Chapter 1 gives an overview of the geotechnical problems in various regions/countries of the world and a general description of the need for ground improvement. A table showing various methods of ground improvement is provided (Table 1.1) that may help the design/site engineer in selecting the best available method.

Chapter 2 describes the field compaction method. A general principle of compaction and field compaction is dealt with in this chapter, along with field specification and control.

Vibro-flotation and dynamic compaction are described in Chapter 3. Design methods and construction procedures are also provided.

Chapter 4 details the replacement method, stage construction, preloading and drainage. Under the drainage method, topics including consolidation, vertical drains, horizontal drains and electro-osmosis are described.

Soil reinforcement is discussed in Chapter 5. In this chapter, natural fibers, geosynthetics, nettings and geomembranes are described.

Chapter 6 describes shallow stabilisation. Various stabilisers such as lime, cement and a few others are discussed.

Deep stabilisation using chemical additives is described in Chapter 7. In this chapter, the details of stabilisation using lime and cement are provided.

Stabilisation using lightweight fills is discussed in Chapter 8.

Chapter 9 details ground improvement using grouting. This chapter details the grouting process using cement and chemicals.

Other techniques for ground improvement, such as pre-wetting, freezing method, geocells, shell footing, biological methods, chemical treatments and protection, alkali activation and carbonation, are described in Chapter 10.

Chapter 11 details site investigation, instrumentation, assessment and control.

## References

Al-Refeai, T. & Al-Ghamdy, D. (1994) Geological and geotechnical aspects of Saudi Arabia. *Geotechnical & Geological Engineering*, 12(4), 253–276.

Azizi, Y., Menani, M. R., Hemila, M. L. & Boumezbeur, A. (2014) Karst sinkholes stability assessment in Cheria area, NE Algeria. *Geotechnical and Geological Engineering*, 32, 363–374.

Christopher, P. M., Jon, C. R., Angela, M. D., Brad, M. B., Craig, R. E., Jackson, Z. L., Jeffrey, P. C., Marisa, H. M., Adrian, A.L.C., Bennett, K., Meshgan, A. & Asma, A. (2016) An unusual inverted saline microbial mat community in an Interdune Sabkha in the Rub' al Khali (the Empty Quarter), United Arab Emirates. *PLoS ONE*. doi:10.1371/journal.pone.0150342.

Edil, T. B. (1994) Immediate issues in engineering practise. In: Den Haan et al. (ed) *Proceedings of Conference on Advances in Understanding and Modelling the Mechanical Behaviour of Peat*. Balkema: Rotterdam, Netherlands. pp. 403–444.

FAO (2019) *Food and Agriculture Organization of the United Nations*. http://www.fao.org/iran/fao-in-iran/iran-at-a-glance/en/ (Retrieved on 11-08-2019).

Fehdi, C., Nouioua, I., Belfar, D., Djabri, L. & Salameh, E. (2014) *Detection of Underground Cavities by Combining Electrical Resistivity Imaging and Ground Penetrating Radar Surveys: A Case Study from Draa Douamis Area (North East of Algeria)*. H2Karst Research in Limestone Hydrogeology. Environmental Earth Sciences, Springer: Cham. pp. 69–82.

Irsyam, M., Susila, E. & Himawan, A. (2007) Slope failure of an embankment on clay shale at km 97 + 500 of the Cipularang toll road and the selected solution. *International Symposium on Geotechnical Engineering, Ground Improvement and Geosynthetics for Human Security and Environmental Preservation, Bangkok, Thailand*. pp. 531–540.

IS 13094 (1992) Selection of ground improvement techniques for foundation in weak soils – Guidelines (Indian Standard).

Muntohar, A. S. (2006) The swelling of expansive subgrade at Wates-Purworejo roadway sta. 8 127. *Civil Engineering Dimension*, 8(2), 106.

Oktaviani, R., Rahardjo, P. P. & Sadisun, I. A. (2018) The clay shale durability behavior of Jatiluhur Formation based on dynamic and static slaking indices. *International Journal of Scientific & Engineering Research*, 9(5), 1266–1281.

Public Works Department. (1976) *Geology of the Republic of Singapore*. Public Works Department, Singapore. p. 79.

Tan, S. L. (1983) Geotechnical properties and laboratory testing of soft soils in Singapore. *International Seminar on Construction Problems in Singapore*, Nanyang Technological Institute, Singapore. pp. TSL1–TSL42.

# Chapter 2

# Earthworks and field compaction

## 2.1 Introduction

Virtually all civil engineering constructions will involve earthwork to begin with in the form of site clearing, cutting, filling and compaction, as illustrated in Figure 2.1. Existing soil on site may not be suitable to support structures such as buildings, roads or embankments that are going to be built on it. In its simplest form, soils on site are at least compacted to their maximum dry densities thus improving their shear strength. In cases where the in situ soils are found to be unsuitable, they need to be removed and replaced with better quality imported soils. This imported soil, however, also needs to be properly compacted to enable it to support the structure's load.

In this chapter, field compaction using equipment such as a smooth wheel roller, sheep foot roller and pneumatic rubber tire roller will be described. These methods are commonly used in construction of road base and sub-base. For a thick layer of granular soil or deep soft compressible soil, compaction methods such as vibro-flotation and dynamic consolidation may be considered. These methods will be described in Chapter 3 of this book.

Sometimes the properties of the in situ soils can be improved by adding suitable additives. In this chapter we will discuss methods of shallow stabilisation using chemical additives such as cement, lime and other admixtures. Such chemicals may also be used to stabilise thick deposits of soil, especially soft soil. This is better known as deep stabilisation, and will be discussed in further detail in Chapter 7 of this book.

## 2.2 General principles of compaction

Compaction can be described as densification of soil using external compression forces (static or dynamic) in order to expel air from the soil voids, while the volume of solids and the water content remain essentially the same. The benefits of compaction can be listed as follows:

- Increase in the soil shear strength
- Increase in unit weight or dry density of the soil
- Reduce soil permeability
- Reduce soil settlement, especially if the soil is subjected to dynamic load such as traffic load
- Control further change in soil moisture content.

The general principle of compaction can be described as follows:

> When water is added to soil with low moisture content, it will act as "lubricant" between the soil particles. This enables the soil particles to come closer and be compacted easily. When water is added continually to the soil, its lubricating effect will also be increased as it covers larger number of soil particles, but in contrast with the ability of the soil to absorb moisture. This will cause the soil voids to increase, and results in the soil dry density to reduce. Therefore, for a particular type of soil, and for a particular type and level of compaction applied, there will be an optimum moisture content that yields the densest state of the soil particles. This condition of the soil is known as the "maximum dry density".

An optimisation of the soil densification process can be carried out by adopting the following steps:

1. The soil placement conditions: its water content, density, depth of layer, etc.
2. Appropriate equipment use, such as roller, vibro-compactor, tamping is selected and followed by the method of operations such as number of passes of rollers, patterns of tamping (in dynamic compaction), etc.
3. Adequate quality control procedure, type and number of tests, statistical evaluation and so forth is to be set up.

## 2.3   Laboratory compaction test

Laboratory tests are usually carried out to simulate the field compaction procedure and to determine the maximum dry density and optimum moisture content for various types of soils. The tests carried out in the laboratory are (1) the light compaction test (2.5 kg rammer method), and (2) the heavy compaction test (4.5 kg rammer method or modified American Association of State Highway and Transportation Officials (AASHTO) method).

Figure 2.1 shows a typical plot of dry density ($\gamma_d$) versus moisture content, w(%). From this figure, the following can be observed:

- The maximum dry density and optimum moisture content of a soil is dependent on the level of compaction energy applied
- If a large compaction energy is applied, then the maximum dry density is also high
- The higher the compaction energy the lower optimum moisture content.

From the results of the laboratory compaction test, a specification for field compaction can be drawn. In many instances, the agency in charge of compaction is expected to achieve a relative compaction (RC) of 90% or more based on the laboratory tests results. Relative compaction is defined as:

$$RC = \frac{\gamma_{d\,(field)}}{\gamma_{d\,(laboratory)}} \qquad (2.1)$$

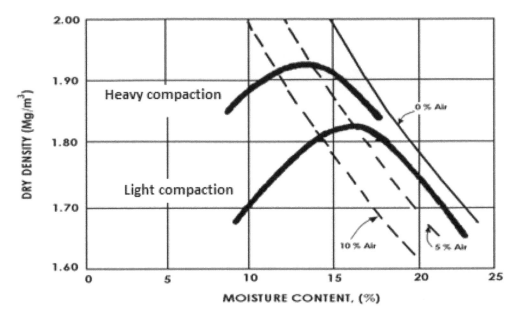

*Figure 2.1* A typical laboratory compaction test result

## 2.4    Field compaction

Field compactions are usually carried out using equipment or machinery utilising static pressure. These are either self-propelled or need to be driven. Further, field compaction using dynamic force is discussed in Chapter 3.

The equipment normally utilised for field compaction includes:

1. Smooth wheel roller
2. Pneumatic tire roller
3. Sheep foot roller
4. Grid roller
5. Vibrating roller
6. Impact roller.

### 2.4.1    Smooth wheel roller

The smooth wheel roller (Figure 2.2) comprises a steel drum whose mass could be increased by adding water or sand ballast. The roller is suitable for most types of soils, especially where crushing action is needed, except for silty and uniform sand.

### 2.4.2    Pneumatic tired roller

The pneumatic tired roller (Figure 2.3) is suitable for most types of fine and coarse-grained soils but not for uniformly graded soils. Increasing the tire inflation pressure could increase the compactive effort. The compactive effort depends on (1) gross weight, (2) wheel

*Figure 2.2* Smooth wheel roller

Source: www.theconstructor.org

*Figure 2.3* Pneumatic tired roller

Source: www.directindustry.com

diameter, (3) wheel load, (4) tire width and size, and (5) inflation pressure. The maximum depth of soil that could be compacted is about 250–300 mm.

### 2.4.3　Sheep foot roller

The sheep foot roller (Figure 2.4) comprises a hollow steel drum with numerous feet in the form of clubs protruding from its surface. The feet will exert relatively high pressure over a small surface. Generally, the wetter and softer the soil, the larger is the contact area required

*Figure 2.4* Sheep foot roller
Source: www.engineeringintro.com

for its optimum compaction. The sheep foot roller is more suitable for fine-grained soils, especially those with water content on the dry side of the optimum. This roller is also suitable for coarse-grained soils with over 20% fines. Moisture control is made easier because of the pockmarked surface during compaction. Steel rollers may be used to level off areas worked by sheep foot or rubber-tired rollers.

### 2.4.4  Grid roller

The grid roller (Figure 2.5) comprises a network of steel bars in the form of heavy square grids. This creates high contact pressures while preventing excessive shear deformation, which is responsible for plastic wave ahead of the roll. Kentledge can be added to the roller to increase its weight. Therefore, these rollers are only suitable for most fine-grained soils, as they break or rearrange gravel or cobble-sized particles.

### 2.4.5  Vibrating roller

The vibrating roller is actually a smooth wheel roller fitted with a vibrator (Figure 2.6). This roller is suitable for most types of soils, particularly effective for those soils whose moisture contents are slightly on the wet side of optimum. The roller is most suitable for

*Figure 2.5* Grid roller

Source: www.gumtree.co.za

*Figure 2.6* Vibrating roller

Source: www.engineeringintro.com

coarse-grained soils with little or no fine. The lightweight vibratory rollers are not of much use as their vibrational amplitude is only of the order of 1–2 mm.

Lewis (1966) made a comparison of the effectiveness of various types of rollers, as shown in Table 2.1. Figure 2.7 shows the effect of various types of compaction equipment on silty clay based on the study of William and Maclean (1950). The compaction curves obtained were for field compaction with 64 passes.

The compactive effort of vibrating rollers is primarily dependent on:

1. Static weight
2. Frequency and amplitude
3. Roller speed
4. Ratio between frame mass and drum mass
5. Drum diameter
6. Impact roller.

Table 2.1 A comparison of the effectiveness of various types of rollers (Lewis, 1966)

| Type of equipment | Average of result in compaction of equipment | | | | | | |
|---|---|---|---|---|---|---|---|
| | Compacted width (mm) | Rotation velocity (m/min) | Number of passes | Area of compaction per hour (m²) | Depth of layer (mm) | Result of compaction per hour (m³) | |
| Smooth wheel rollers | 1800 | 70 | 4 | 1220 | 150 | 185 | Suitable for all types of soil except silty and uniform grade sand |
| Vibratory rollers | 2000 | 37 | 4 | 870 | 300 | 265 | Suitable for all types of soil |
| Pneumatic tired rollers | 2400 | 66 | 3 | 4000 | 250 | 612 | Suitable for all types of soil but not for uniformly graded soils |
| Sheep foot rollers | 3700 | 270 | 6<br>14<br>32 | 8200<br>3500<br>1530 | 225 | 1875<br>804<br>350 | Different passing number needed for clay, sandy clay and gravel/sand |
| Grid rollers (80 hp) | 1600 | 135 | 7 | 1500 | 200 | 300 | Suitable for all type of soil with wide range of water content |
| Grid rollers (150 hp) | 1600 | 270 | 8 | 2640 | 200 | 536 | Not suitable for uniformly graded sand/wet soil |
| Vibratory rollers (Stothert & Pitt, 72,000 kg) | 1700 | 40 | 7 | 485 | 225 | 111 | Suitable for granular soils |

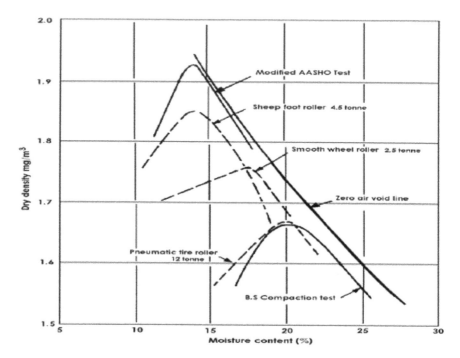

*Figure 2.7* Effect of various types of compaction equipment on silty clay

Source: After William and Maclean (1950)

*Table 2.2* Typical characteristics of impact and vibratory rollers

| Type of roller | Mass (t) | Max working speed (km/h) | Vibrating frequency (Hz) | Depth of lift (m) | Number of passes |
|---|---|---|---|---|---|
| Vibrating rammer | 0.1–0.3 | 0.6–0.8 | 7–10 | 0.2–0.4 | 2–4 |
| Light vibrating plate | 0.06–0.8 | I | 10–80 | 0.15–0.5 | 2–4 |
| Light vibrating roller | 0.2–0.6 | 2–4 | 25–70 | 0.3–0.5 | 4–6 |
| Heavy vibrating roller | 6–15 | 8–10 | 25–30 | 0.3–1.5 | 4–6 |
| Heavy self-propelled vibrating roller | 6–15 | 6–13 | 25–40 | 0.3–1.5 | 4–6 |
| Impact roller | 7 | 10–14 | – | 0.5–3 | Up to 30 |

Impact roller consists of a non-circular mass which is towed along the ground. As its center of gravity rises and falls, its mass exerts a high impact force causing compaction of the soil. Clifford (1978) described a new type of impact roller developed in South Africa consisting of a 1.5 m thick "square" roller with rounded edges. It was found suitable for natural ground and filled-up soil. Because the impact roller leaves the surface uneven (small depressions), it is recommended for sub-grades and earth fills rather than for surface works. The typical characteristics of impact and vibratory rollers are presented in Table 2.2.

In general, the equipment compacts the soil using the following three methods:

- *Pressure*: The contact pressure between the equipment and the soil is the most important factor that results in compaction of the soil below. For example, a sheep foot roller can exert a contact pressure as high as 3500 kPa.
- *Vibration*: The vibrating roller uses vibration to increase the compaction of soil particles. Vibrators normally use a vibration range of 1000–3500 cycles per minute.
- *Manipulation*: Compaction equipment also applies shear stresses to soil to assist the compaction. This action is known as manipulation. Excessive manipulation on a rather damp earth fill can result in adverse effect, however (i.e., a reduction in compaction). In a case like this, the soil could not be compacted but will experience a condition known as "pumping." Pneumatic tire and sheep foot roller in general compact soil with pressure and surface manipulation or kneading.

## 2.5   Field specification and control

The method used for specification and control of field compaction will depend on the condition of the site. In cases where the work is confined to a small area, such as a building or a commercial complex, most likely the soil is quite uniform. In this case, the compaction operation can be easily controlled. Specifications are then made in terms of operational methods, such as the maximum thickness of each soil layer to be compacted, the weight and type of compaction equipment to be used and the number of passes for each layer. In larger work, for example dam construction, the operational method is usually determined after a field trial has been made on the actual soil and using the same equipment.

On the site where the soil is more diverse, for example in road construction, it is usually better to set the desired results that are to be obtained and not the operational methods – for example, setting the desired value of relative compaction, as was discussed earlier. Table 2.3 shows the typical relative compaction expected in various engineering projects.

The following four methods are generally used to check in situ density of soil as per availability:

1. Sand replacement method
2. Core cutter
3. Rubber balloon
4. Nuclear density method.

*Table 2.3* Specification of relative compaction

| Types of Project | Minimum Relative Compaction Required (%) |
|---|---|
| Fill for buildings and roads | 90 |
| Sub-grade – top 150 mm below road | 95 |
| Aggregate base below road | 95 |
| Earth dam | 100 |

The parameters of soil to be measured are the weight of the soil (W), the volume of the soil (V) and the moisture content (w). Soil bulk density ($\gamma_b$) can be calculated as

$$\gamma_b = W/V \tag{2.2}$$

And dry density of soil, $\gamma_d$

$$\gamma_d = \gamma_b / (1 + w) \tag{2.3}$$

### 2.5.1    Sand replacement method

In the sand replacement method, a hole is dug into the ground. The weight and moisture content of the excavated soil is determined. The volume of the excavated hole is measured by filling the hole with sand of known bulk density poured from a cylinder. The error in this method could result from collapse of the hole sides as well as changes in density of the sand due to changes in moisture content. However, this method usually gives a satisfactory result if conducted with great care. The arrangement of the test equipment is shown in Figure 2.8.

### 2.5.2    Core cutter

The core cutter test (see Figure 2.9) is performed by driving a core cutter of known volume into the ground and then retrieving it together with the soil inside. The weight and moisture content of the soil sample is determined to calculate the soil density. This method is suitable for fine-grained soil with no coarse materials (sand or gravel).

### 2.5.3    Rubber balloon method

The rubber balloon method requires a hole to be excavated into the ground (Figure 2.10). The volume of the hole is determined by measuring the quantity of water required to fill up the balloon. However, if the wall of the hole is too rough, error may be introduced in the

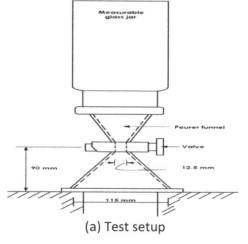

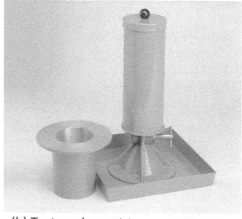

(a) Test setup     (b) Test equipment (source: civildigital.com)

*Figure 2.8* Sand replacement method

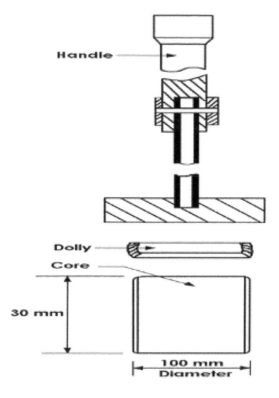

Figure 2.9  Core cutter

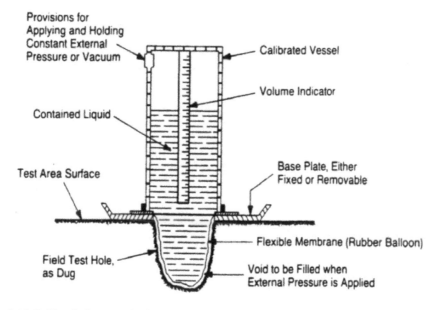

Figure 2.10  Rubber balloon method

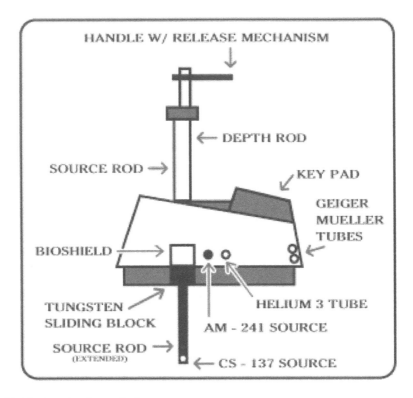

*Figure 2.11* Nuclear density method

Source: www.apnga.com

measured volume of the hole, as the balloon will try to bridge any irregularities of the hole's wall. The soil density is calculated by measuring the volume of the hole as well as the weight and moisture content of the displaced soil.

### 2.5.4 Nuclear density method

There are two basic methods normally used, that is back scattering and direct transmission. This method however, requires skilled personnel and extensive calibration of the equipment (Figure 2.11).

## References

Clifford, J. M. (1978, December) The impact roller - problems solved. *The Civil Engineer in South Africa*, 20(12), 321–324.

Lewis, W. A. (1966) Full-scale studies of the performance of plant in the compaction of soils and granular base materials. *Proceedings of the Institution of Mechanical Engineers: Automobile Division*, https://doi.org/10.1243/PIME_AUTO_1966_181_015_02

Williams, F.H.P. & Maclean, D. J. (1950) *The Compaction of Soil*. Road Research Technical Paper No 17. Department of Scientific and Industrial Research, HMSO London.

# Chapter 3

# Vibro-flotation and dynamic compaction

## 3.1 Introduction

One of the most common ground improvement techniques uses vibration. With this technique, the density of granular soils is increased by the insertion of a heavy vibrating poker to a desired depth. This vibrating poker is also known as a depth vibrator, a vibrofloat or simply a float. Water flushing is done during the insertion process. Vibro-flotation is most suitable for very loose sands submerged under the water table. The efficiency of densification reduces with an increase in silt and clay content, and the penetration rate reduces with dense sands and a deep water table.

Dynamic compaction is a ground improvement technique that densifies soils and fills by using a drop weight. The drop weight – typically hardened steel plates – is lifted by a crane and repeatedly dropped on the ground surface. The drop locations are typically located on a grid pattern, the spacing of which is determined by the subsurface conditions, foundation loading and geometry. Treated granular soils and fills have increased density, friction angle and stiffness. The technique has been used to increase bearing capacity and decrease settlement and liquefaction potential for planned structures. It has also been used to compact landfills prior to construction of parking lots and roadways, and to stabilise large areas for embankment works.

## 3.2 Vibro-flotation

The procedure using a vibrating probe to densify loose granular soil was invented in Germany in the 1930s. Nowadays, the main application of this technique is to strengthen cohesive soil by inserting sand or stone into holes formed by the vibrator. The holes are formed by the combined weight of the vibrator, its vibration and the jetting action of air or water. This method is known as the stone column method or vibro-replacement. In addition to the stone column method, vibro-flotation can also be used to densify granular soil such as sand. This technique is known as vibro-compaction, sand compaction or sand column.

In both methods, the shear strength of the soil will be improved. The total and differential settlement can be reduced. In the sand compaction method, sand with relative density that is initially loose will be densified, thereby giving good protection to the soil against liquefaction during dynamic loading, such as during an earthquake. The stone column method, besides strengthening soft cohesive soil, will also act as a vertical drain to allow for quick dissipation of excess pore water pressure due to its high permeability.

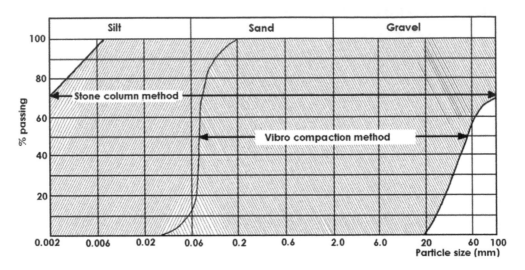

*Figure 3.1* Suitability of soil for vibro-flotation

Particle size distribution of the soil may be used as a guide as to which method is to be used – sand compaction (vibro-compaction) or stone column (vibro-replacement) (see Figure 3.1). Soil with a silt content of less than 15% can be densified with the sand compaction method (vibro-compaction), while clay and clayey soils can be treated with the stone column method (vibro-replacement).

### 3.2.1   Design principle

#### 3.2.1.1   Vibro-compaction (sand compaction or sand column)

The engineering properties of granular soil (its permeability, shear resistance and resistance toward dynamic load) are very dependent on the relative density of the said soil.

High relative strength will give equally high safe bearing capacity and low settlement. In the case of seismic loading, it is known that soil resistance to liquefaction is a function of the relative density of the soil, while in the problem of soil retention, the earth active pressure reduces while the passive pressure increases when the relative density (and the shear strength) of the soil increases.

In vibro-compaction work, the compaction points are usually arranged in a triangular pattern (center to center in plan) with sufficient depth to densify the required volume of granular soil.

Figure 3.2 shows the percentage of relative density that may be achieved versus the area for each of the treatment centers. In practice, it is unlikely that we will achieve relative density of greater than 85%, while a value of 55% is the lower limit.

The effectiveness of this vibro-compaction technique for granular soil can be tested with an in situ test like the standard penetration test (SPT), cone penetration test (CPT) or plate loading test.

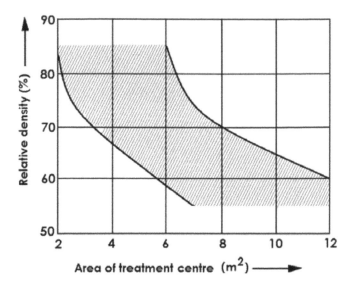

*Figure 3.2* Percentage of relative density that may be achieved versus area for each of the treatment centers

### 3.2.1.2   Stone columns (vibro-replacement)

As mentioned earlier, soft soils can be treated with the stone column or vibro-replacement technique. In this technique, replacing part of the soil with the compacted stones treats the soil. The stone columns formed are usually about 0.9 m in diameter. The typical depth of soil treated is about 10 m.

Estimates on the distance between columns and its effect on settlement is normally based on a semi-empirical method using bearing capacity theory, taking into account the passive column wall resistance (Bell, 1975). Nevertheless, this method does not take into account the possibility of improvement in the subsoil due to compaction or drainage during installation of the columns.

The stone column method is most successful at forming columns in normally consolidated soft clays, silt and a thin layer of peat (Huat and Ali, 1992, Huat et al., 1993b). Construction has to be done to allow dissipation of excess pore water pressure at each stage to allow for sufficient gain in shear strength of the subsoil. According to Bell (1975), this method enables stone columns to be formed in soil with undrained shear strength as low as 10 kN/m². 

In Japan, a variant of the sand columns, better known as sand compaction piles, are generally used. As its name implies, sand backfill rather than gravel is used.

In cases where the upper zones of the in situ soil do not have adequate strength to provide lateral support to the sand or stone columns, synthetic fabrics (geotextile) wrap has been suggested to remedy the problem (Al-Refeai, 1992).

Another version of the column technique is what is known as the vibrated concrete column. In this technique the concrete columns are created in situ. A vibrator penetrates the weak subsoil until it reaches the proposed bearing stratum. Concrete is then pumped as the vibrator is withdrawn. By revibrating the concrete at the base and top, bulbous ends are

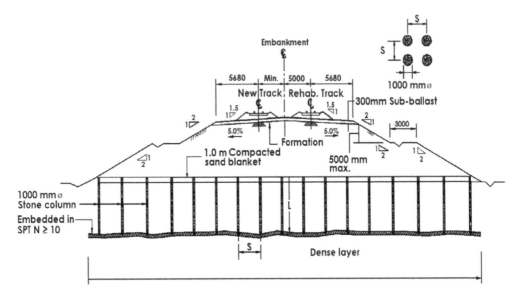

*Figure 3.3* Cross-section of an embankment on stone columns

formed which enhance base resistance and form pile caps at the surface. The embankment above the piles is reinforced with geogrid layers to promote arching and load transfer to the columns. An example of this application is given by Maddison et al. (1996).

The design and construction of both the vibro-compaction and stone column (vibro-replacement) methods requires specialist knowledge. Because the subsoil condition may become very complex during and after the treatment, design assumptions have to be checked by initial trial or by conducting relevant tests during the early stage of column installation.

Arulrajah and Abdullah (2002) describe the design and performance of the stone column method for construction of a high-speed railway embankment on soft ground. Figure 3.3 shows a typical cross-section of the embankment.

### 3.2.2 Construction method

#### 3.2.2.1 Vibro-compaction (sand compaction or sand column)

The construction procedures of the vibro-compaction method are shown in Figure 3.4.

1.  The vibrator suspended from a crane is positioned on the selected treatment point. With the aid of a down jet and self-weight of the vibrator, the vibrator is inserted into the ground to the desired depth, usually about 5 m.
2.  The water flow is then changed to an upward jet and compaction begins. Vibration applied to the soil rearranges the soil particles. Sand and gravel is then poured into the hole with the aid of water flow.
3.  The vibrator is then lifted in stages, forming a compacted soil cylinder (i.e., influence zone) of diameter 2–4 m in the process. The typical spacing treatment range is between 1.5 m and 4 m.

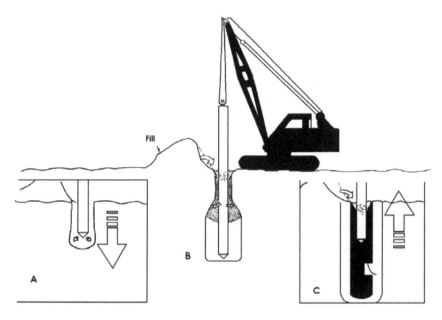

*Figure 3.4* The construction procedures of the vibro-compaction method

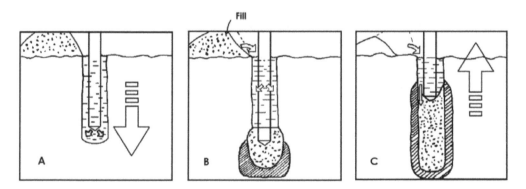

*Figure 3.5* The construction procedure of stone column

### 3.2.2.2   Stone columns (vibro-replacement)

This technique is used to treat cohesive soil like clay and silt. The construction procedure is shown in Figure 3.5.

1.  The hole is first formed in the ground with the combined weight of the vibrator, vibration and jetting action of water or air.
2.  The vibrator is then withdrawn. Gravel or crushed stone is then poured into the hole. The vibrator is then reinserted to compact those fill materials. Radial forces applied by the vibrator will force the gravel or crushed stone to be displaced sideways into the ground.

3. The vibrator is then withdrawn in stages. These cycles of pouring gravel or crushed stone into the hole and compacting it are repeated until the design limit is achieved.
4. The columns are usually formed in a grid in the plan area. Typical spacing of the grid is about 1.0–2.5 m. Typical diameter of the vibrator is 0.3–0.4 m.

Machinery for installation of the columns usually requires a working platform of good quality granular fill material to ensure a smooth operation. Water of about 35 m³/hr is required on site. Water with fines suspended in it needs to be subsequently discharged from the site.

In this technique, the weak soil is partially replaced and displaced by the introduction of these stiffer reinforcing elements at regular grid pattern intervals, and the response of this modified ground becomes complex. There are ways to arrive at an equivalent stiffness matrix of a system that replaces some part with a material of larger stiffness.

Similarly, there are ways to establish the modified density and stiffness when the entire soil mass is densified. When the improvement is attributed to both displacement and replacement, the quantification of improvement is difficult to determine. Considerable efforts like large-scale load tests can only prove the effectiveness of the installed stone columns.

In a first step, an improvement factor is established by which stone columns improve the performance of the subsoil in comparison to the state without columns just by increasing the overall stiffness. The grid patterns and concept of the unit cell are illustrated in Figure 3.6. A basic improvement factor can be arrived at based on the area replacement ratio and the reinforcing material used for stone columns.

The improvement factor is presented in Figure 3.7. According to this improvement factor, the deformation modulus of the composite system can be established due to which settlements will be reduced. Priebe's method is a unit cell approach, which takes into account oedometric conditions. This is very important because the direct use of Priebe's composite parameters for slope stability results in an unconservative safety factor.

The deformation modulus of the composite system is one of the basic inputs for finalising the design of stone columns. However, the reality is that in many practical cases,

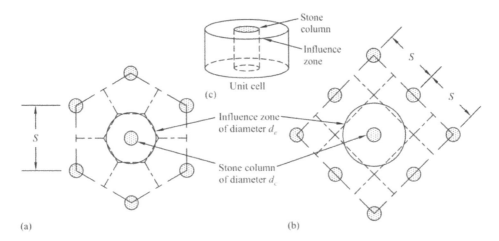

*Figure 3.6* Unit cell concept: (a) influence zone in triangular pattern ($d_e$ = 1.05 S); (b) influence zone in square pattern ($d_e$ = 1.13 S); (c) unit cell (with permission from ASCE)

Source: After Das and Deb (2017)

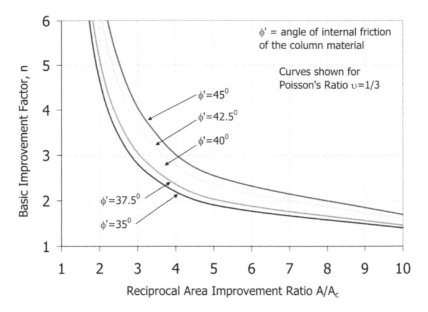

*Figure 3.7* Priebe's basic improvement factor

Source: Reproduced from Priebe (1995)

the reinforcing effect of stone columns installed by vibro-replacement is superposed with the densifying effect of vibro-compaction, that is, the installation of stone columns densi-fies the soil between grids, increasing its k0 and kp. In such a case, the densification of the soil has to be evaluated on the basis of original soil data, and correspondingly the design of vibro-replacement can be modified to suit a particular improved site condition.

The basic improvement factor ($n_0$) shall be calculated using the formula

$$n_0 = 1 + \frac{A_c}{A} \left[ \frac{5 - A_c/A}{4 K_{aC} \left(1 - A_c/A\right)} - 1 \right]$$

Since the column cannot fail in end bearing and any settlement of the load area results in a bulging of the column, which remains constant throughout its length, the following two corrections need to be applied to obtain an appropriate value for the improvement factor:

- Correction for column compressibility
- Correction for overburden.

### 3.2.3  Correction for column compressibility

In the case of soil replacement, the actual improvement factor does not achieve an infinite value as determined theoretically for non-compressible material, but it coincides at best with the ratio of the constrained moduli of column material and soil. Due to the compressibility

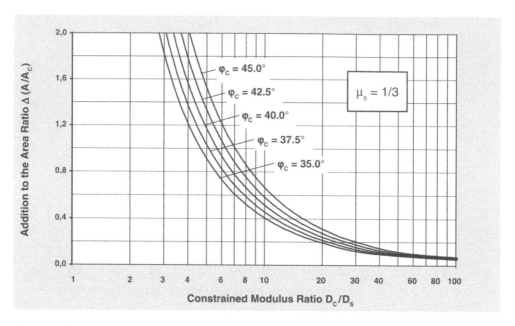

*Figure 3.8* Consideration of column compressibility

of column material, the area of the column may increase and the improvement factor will be reduced. The improvement factor after the compressibility correction can be calculated using the following relation and Figure 3.8.

$$n_1 = 1 + \frac{A_c}{A}\left[\frac{\frac{1}{2} + f(\mu_s . A_c / A)}{K_{aC} . f(\mu_s . A_c / A)} - 1\right]$$

### 3.2.4   Correction for overburden

As a result of column installation, the weights of the columns WC and of the soil WS, which possibly exceed the external loads considerably, has to be added to the external loads. Without consideration of these additional loads, the initial pressure difference decreases asymptotically and the bulging is reduced correspondingly. In other words, with increasing overburden, the columns are better supported laterally and therefore can provide more bearing capacity.

Since the pressure difference is a linear parameter in the derivations of the improvement factor, the ratio of the initial pressure difference and the one depending on depth expressed as depth factor $f_d$ delivers a value by which the improvement factor $n_1$ increases to the final improvement factor $n_2 = f_d \times n_1$ on account of the overburden pressure.

$$f_d = \frac{1}{1 + \dfrac{K_{0C} - 1}{K_{0C}} . \dfrac{\sum(\gamma_s . \Delta d)}{p_C}}$$

$$n_2 = f_d \cdot n_1$$

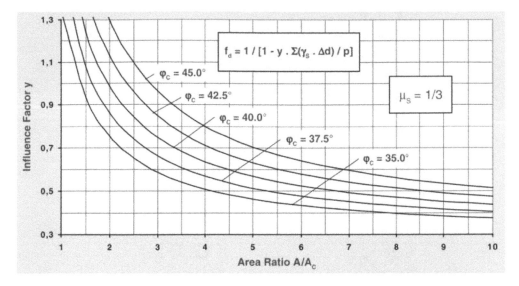

Figure 3.9 Determination of depth factor

The depth factor can be calculated by using Figure 3.9.

## 3.3  Miscellaneous techniques

### 3.3.1  Geopiers

Geopiers (or preload piers) are another example of a floating foundation technique for stabilising soft soil, including peat (Wissmann et al., 2002; Biringen and Edil, 2003). These piers are inserted in the ground and expanded to about three times their original cross-sectional area. As a result of this expansion, the surrounding soil is radially stressed, which initiates consolidation and general stiffening of the surrounding soil. If needed, vertical wick drains inserted between the preloaded piers can accelerate consolidation. The end result is a pier-soil composite with enhanced stiffness and strength to support foundation loads within tolerable settlement.

This process can be viewed as a cavity expansion problem. Figure 3.10 shows the Geopier construction. Figure 3.11 shows a comparative performance of a Geopier and stone column.

### 3.3.2  Mortar sand rammed column (MSRC)

The MSRC is a variant of the floating pile foundation that combines the concept of a vertical drain (Figure 3.12). The vertical drain component is the sand surrounding the mortar, and the whole system is a floating pile foundation. Theoretically, the ramming process increases the matrix soil lateral earth pressure in the vicinity of the columns and between the columns, thus making the soil stiffer. In time, as excess pore water pressure dissipates to the nearest column, the compressibility as well as shear strength of the underlying soil will improve.

These columns can be constructed in a five-step process, shown in Figure 3.13. Holes of 0.5–1.0 m diameter are drilled to a depth that typically varies from approximately 2 m to

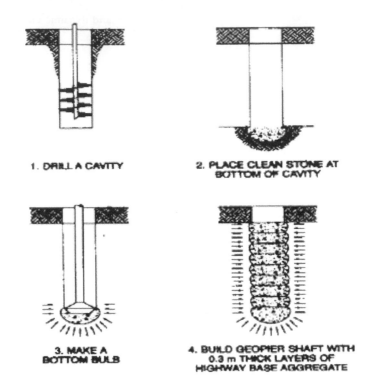

1. DRILL A CAVITY

2. PLACE CLEAN STONE AT BOTTOM OF CAVITY

3. MAKE A BOTTOM BULB

4. BUILD GEOPIER SHAFT WITH 0.3 m THICK LAYERS OF HIGHWAY BASE AGGREGATE

*Figure 3.10* Geopier construction
Source: Kwong et al. (2002)

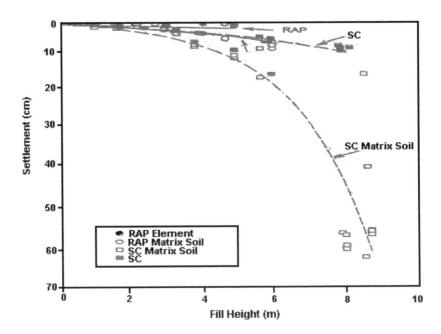

*Figure 3.11* Comparison of rammed aggregate pier (Geopier) and stone column
Source: Edil (2003)

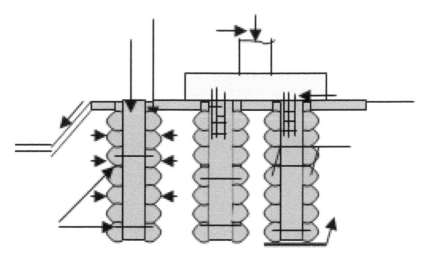

*Figure 3.12* Concept of MRSC
Source: Sulaeman (2003)

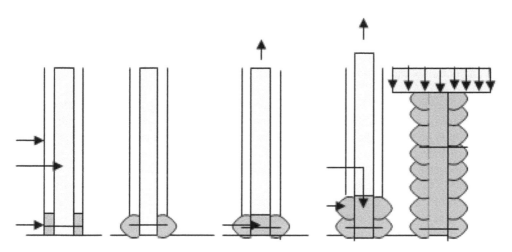

*Figure 3.13* MRSC construction
Source: Sulaeman (2003)

10 m below the ground surface; a temporary casing should be employed to separate mortar and compacted sand. Ramming a layer sand with a high-energy tamper to form undulating, very dense sand stabilises the bottom layer of sand. The process should be carried out layer by layer to avoid difficulties in withdrawing the casing. The subsequent step is the preloading and waiting period; the duration is based on a calculation of the soil parameter and prescribed performance. The preloading further prestresses and restrains the columns and the surrounding soil matrix to increase the stiffness and strength of this foundation system (Sulaeman, 2003).

## 3.4   Dynamic compaction

Dynamic compaction (also known as dynamic consolidation) is a treatment technique suitable for most types of soils, including very soft soil such as peat and organic silt, where large displacement may occur. It consists of two steps: area-wide compaction and an ironing pass. In this technique, the soil is compacted by tamping the soil surface (Menard and Broise, 1975).

This treatment technique can be used for the following types of soil:

1.   Rock fill
2.   Loose sand and alluvial gravel
3.   Non-cohesive earth fill
4.   Sandy and silty clay, silt (with or without preloading and vertical drains).

The main objectives of this technique are as follows:

1.   To improve soil bearing capacity
2.   To reduce total and differential settlement
3.   To reduce the danger from liquefaction, particularly in earthquake-prone areas.

### 3.4.1   Design

The dynamic compaction technique involves the dropping of weights (10–20 tons) repeatedly onto the soil surface from a height of 20–40 m. Each drop generates a system of transient and complex waves (as shown in Figure 3.14):

• An expansive wave (also known as a main wave or P-wave) with a typical velocity of 3000 m/s travels in the liquid phase of the ground. This wave creates a push-pull deformation in pore water that results in an increase in pore pressure as well as destruction in the soil structure.
• A transverse wave (also known as an S-wave), which is slower than the P-wave and travels in the solid phase of the ground.
• A surface wave (also known as a Rayleigh wave), with vertical and horizontal components that travel along a cylindrically shaped front.

Both S- and P-waves will cause the soil grains to move in a horizontal direction. Due to this movement, the soil grains will be rearranged in a denser condition, thereby reducing the soil voids. The danger toward liquefaction is also thereby reduced. A suitable drainage path is also formed inside the soil.

The energy is generally applied in phases on a grid pattern over the entire area using either single or multiple passes. Following each pass, the craters are either leveled with a bulldozer or filled with granular material before the next pass of energy is applied (Lukas, 1995).

Methods to determine the required degree of compaction are mostly empirical and highly dependent on past experience. Therefore this technique is more suited to be performed by specialist contractors.

Rammer weight (M) and height of drop (h) depends on the thickness of the layer (H) to be compacted. The energy produced by each drop, Mh, is an important design parameter. This

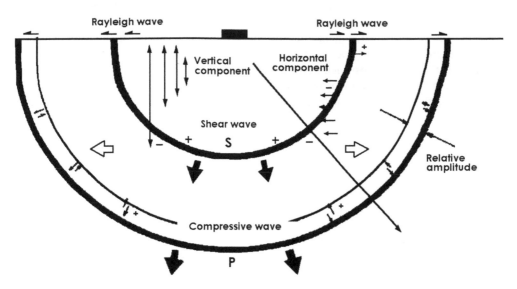

*Figure 3.14* Types of waves and their direction of travel

usually ranges from 1500 kNm to 5000 kNm and can be as high as 20,000 kNm. The tamping energy can be approximately related to the layer thickness as follows:

$$Mh > 10\,H^2 \tag{3.1}$$

where h and H are expressed in meters, and M in kN.

The depth of improvement generally depends on the total amount of energy applied to the soil, which is a function of the weight of the rammer and the drop height. It can be calculated from Equation (3.2).

$$D = n\sqrt{WH} \tag{3.2}$$

where

  $D$ = anticipated depth of treatment
  $n$ = a constant between 0.3 and 0.6
  $W$ = weight of the rammer
  $H$ = height of drop.

The ironing pass consists of dropping a lighter rammer (~4 tons) with smaller drop heights (5–7 m) on an overlapping grid pattern. The purpose of the ironing pass is to compact the soils at shallower depths, up to the depths of the craters. The size of the rammer (in plan) for the ironing pass is generally larger than that used for deep compaction.

The final design of the system needs to take into account the magnitude and shape of the weight (rammer), the height of drop, the frequency of drop and the distance between the ramming (dropping) points. Usually a preliminary compaction needs to be done on a test section with a comprehensive program of laboratory and in situ tests.

The soil response to the treatment can be monitored by using various methods, such as by monitoring settlement, standpipes to observed pore water pressure and total stress sensor ($k_o$ monitor), while an in situ test such as a standard penetration test (SPT) or cone penetration test (CPT) can be performed to measure the strength parameter of the treated soil.

Figure 3.15 shows a typical site treated with the dynamic compaction method. Figure 3.16 and Table 3.1 show the typical results of soil treated with this technique.

*Figure 3.15* A typical site treated with the dynamic compaction method

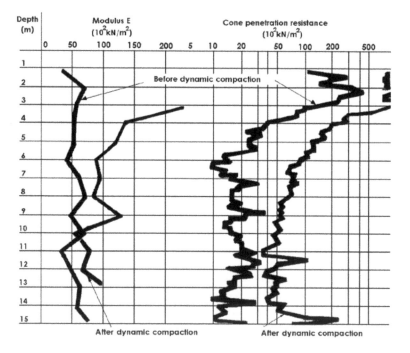

*Figure 3.16* A typical result of treated soil

Table 3.1  Typical results of treated soil ('dynamic consolidation', www. soletanchefreyssinet.com)

| Soil Type | Improved Bearing Capacity (%) | Final Allowable Bearing Pressure (kN/m²) |
|-----------|-------------------------------|-------------------------------------------|
| Clayey soil | 100–150 | 100 |
| Silt | 200 | 200 |
| Sand | 400 | 350 |

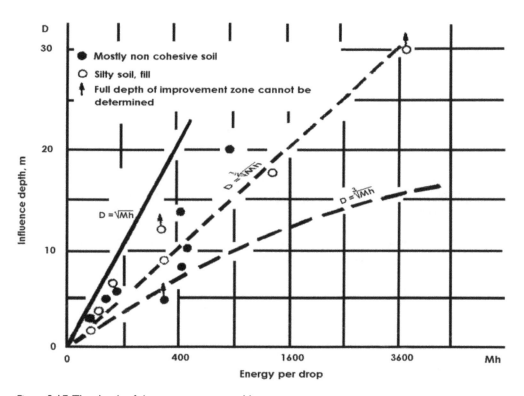

Figure 3.17 The depth of the treatment versus blow energy

Figure 3.16 shows the performance of a treated soil versus time. It can be seen that the soil strength increases with dissipation of excess pore water pressure following treatment with dynamic compaction. Thus dynamic compactions do give instant results. Settlement occurs during the compaction process by reducing soil volume between 3% and 10%.

The depth of the treatment can be related to the energy created by one drop (or blow), as shown in Figure 3.17.

The dynamic compaction method has a number of limitations, however. For example, when the initial strength of the soil is too low, as in organic soil, the soil cannot be sufficiently improved to the extent that it can support heavy or sensitive structures. In such

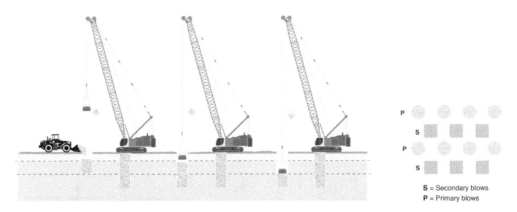

*Figure 3.18* Dynamic replacement to form sand columns

Source: www.trevispa.com

a case, the dynamic compaction can be carried out in combination with other techniques such as:

- Shallow removal of the upper soft soil layers (2–3 m)
- Temporary loading on soil where the compaction is to be carried out
- Vertical drains.

Another variant of the dynamic compaction method is the dynamic replacement (DR) method. The process consists of dropping heavy weights (10–15 tons) from heights of 10–15 m to form large diameter granular columns in cohesive soil deposits. Subject to the site working conditions, the nature and consistency of the soil conditions and the required length of the DR columns, it may be required to create pre-excavated DR columns (Sin, 2003). The size of the excavated crater may be 2–3 m deep over an area of 4 m². Very often, the actual size of this pre-excavated DR crater is governed by the water table condition. After the excavation, the crater is filled with granular materials and a series of pounding and ballasting phases of work shall be carried out until the completion of the installation of a DR column, as shown in Figure 3.18.

### 3.4.2   Construction method

Before starting compaction, the area to be treated is first backfilled with a layer of granular material such as sand of 0.5–1 m thick, to:

- Ensure sufficient bearing capacity for passage of the dynamic compaction machinery
- Prevent the weight drop from penetrating too deep into the treated soil
- Provide drainage to discharge water pressure
- Provide static load (surcharge) of intensity of 10–20 kN/m².

The compaction operation is usually conducted in a series of passages with adequate rest time in between, ranging from a few days to several weeks or months, to allow for sufficient time for excess pore water pressure to dissipate. During the treatment process, the ground condition is monitored to ensure the desired subsoil condition has been achieved.

## 3.5   Case Studies

### Case study 1

M/s Urban Tree Infrastructure Pvt. Ltd. (Urban Tree), Chennai, proposed to develop a residential project in Chennai. The project comprises 198 units of Stilt + 4 floors and the approximate area of development is about 2.5 acres.

The subsoil in the project site comprises desiccated clay and medium dense sand up to about 3.5 m followed by relatively weak clay and sandy clay up to 6 m depth. This top 6 m of soil with highly varying consistency is followed by about 8 m with medium dense sand and stiff clay deposits, after which there is a 6 m thick layer of medium stiff consistency. Denser sand layers and hard clay layers form the remaining subsoil profile. The required loading intensity of the proposed structure on the soft soil is 100 kPa.

Considering the project boundary conditions, the vibro-replacement technique with 20% area replacement ratio (stone columns with dry bottom feed method) up to 6 m depth was adopted as a viable method for subsoil improvement and a full raft foundation supported by the treated ground as an alternative foundation system. A diagram of the soil profile and ground improvement planned is shown in Figure 3.19.

Keeping the importance of the post-construction performance of the structure, a plate load test has been conducted on the improved ground, and also about 14 locations

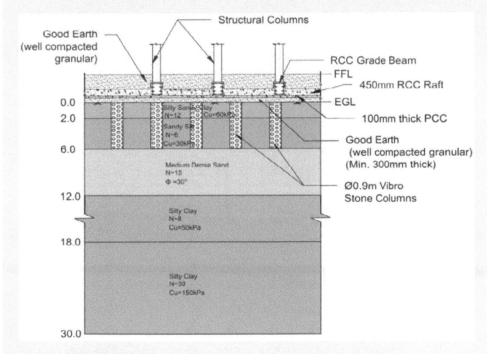

*Figure 3.19* Typical soil profile showing ground improvement arrangement

were identified on the raft foundation to monitor settlements during and after construction.

Test results post-construction are shown in Figure 3.20.

The results of post-construction are shown below:

- Achieved bearing capacity: >150 kPa
- Long-term settlement: <50 mm

The measured settlements are substantially lower than the predicted settlement, which proved the efficiency of the raft foundation resting on improved ground.

A view of the completed building is shown in Figure 3.21.

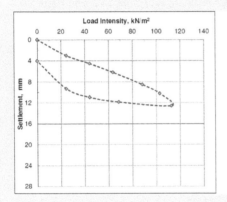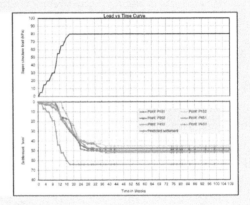

Figure 3.20  Post-construction test results

Figure 3.21  View of completed building

Source: Technical Report on Ground Improvement and Geosynthetics, IGS, New Delhi

## Case study 2

M/s Hindustan Petroleum Corporation Ltd. proposed to develop a tank farm for 18 nos. of floating roof storage tanks. Based on the storage type of the liquid, the tanks are arranged under six enclosures as shown in Figure 3.22. Four (4) nos. of fixed roof storage tank and 3 nos. of fire water tanks. The diameter of a tank is 32 m and the height of a tank is 15 m.

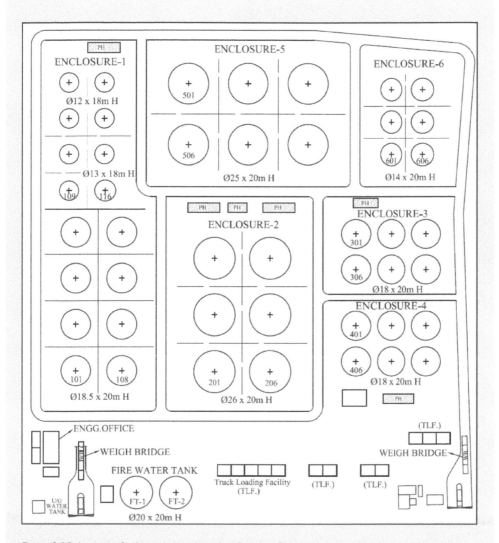

*Figure 3.22* Layout of oil storage terminal at Pipavav, Gujarat

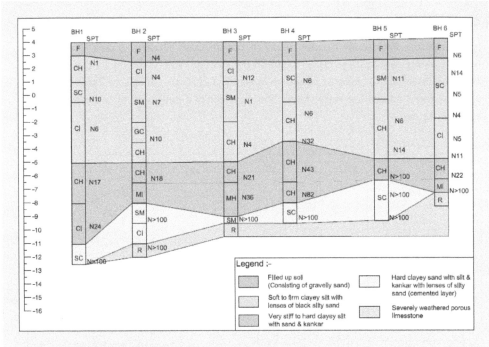

*Figure 3.23* Typical subsoil profile

The subsoil consists of top 2.5 m fly ash fill with the SPT N value zero followed by soft silty clay of SPT N value of 7 up to 9 m depth, and this layer is underlain by medium dense sand of SPT N value of 15 up to 25 m depth (Figure 3.23).

The bearing capacity of virgin soil at the foundation level of the tanks is 100 kPa. The required bearing capacity of a tank foundation is 200 kPa.

To improve the bearing capacity of virgin soil, the vibro-replacement method with a 25% area replacement ratio up to the depth of 11.5 m (stone column using wet top feed method) was proposed.

The typical soil profile and stone column cross-section below the tanks is shown in Figure 3.24.

A schematic diagram of the stone column construction by the top feed method is shown in Figure 3.25, and a photograph during installation is shown in Figure 3.26.

The results of the post-construction tests are as follows:

- Improved bearing capacity: >200 kPa
- Hydro test with full load (water) was carried out and the settlement was recorded. The observed settlement is less than the allowable settlement of each tank (<300 mm) (Figure 3.27).

A completed view of the plant is shown in Figure 3.28.

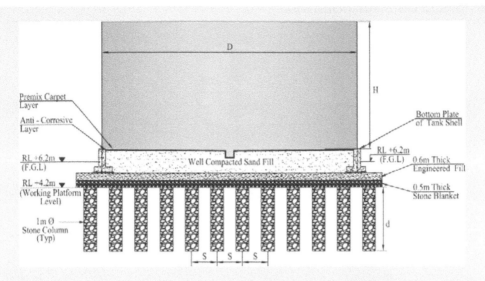

Figure 3.24 Typical soil profile and cross-section of stone column below the tanks

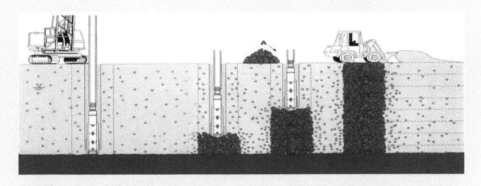

Figure 3.25 Step-wise illustration of vibro stone column installation by top feed wet method

Figure 3.26 Installation of vibro stone columns by top feed wet method

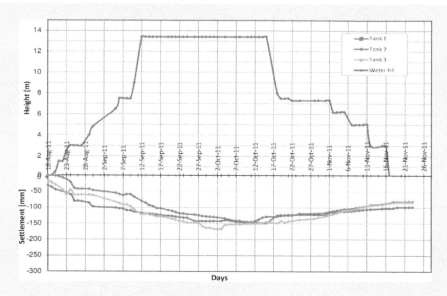

*Figure 3.27* Long-term observed settlement

*Figure 3.28* Completed view of the plant

Source: Technical Report on Ground Improvement and Geosynthetics, IGS, New Delhi

## Case study 3

Ground improvement of an existing ash pond site, Anpara Thermal Power Plant, Uttar Pradesh, India (stone column).

Objective of ground improvement: Expansion of existing thermal power plant.

Site allocated for expansion: An abandoned ash pond of area ~ 5400 acres. Figure 3.29 shows a photograph of the site.

*Figure 3.29* A photograph of the site

Initial soil condition of site:

> Soil strata: Ash deposit 3–13 m; clayey silt/silty clay up to 23 m; dense sandy silt or hard clayey silt with occasionally weathered rock (granitic gneiss)
> State of denseness: Loose to medium dense in condition
> Existing bearing capacity of the fly ash deposit: much less than 100 kN/m².

A typical soil profile is shown in Figure 3.30.

> SPT value of ash deposit: Ranges from N value of 2 to 30, but averages 3 to 8.
> Site falls under Zone III – IS 1893 (Part 1) 1982: Susceptible liquefaction
> Method adopted for improvement of the ash pond: vibro stone column (dry bottom feed method)
> Reason for vibro stone column (dry bottom feed method): This offers a highly economical and sustainable alternative to piling and deep foundation solutions, removing the need to bypass problem ground by densifying and strengthening weak or poorly compacted soils in situ.
> Bottom feed method: Where a high water table or weak soils are present, there is a likelihood of collapse of the borehole when the vibrofloat is withdrawn. Under such conditions, the purpose-built bottom feed system is used to ensure integrity of the stone column from top to bottom. The vibro-rig is fitted with a hopper which feeds stone into a tremie pipe running down the length of the vibrofloat.

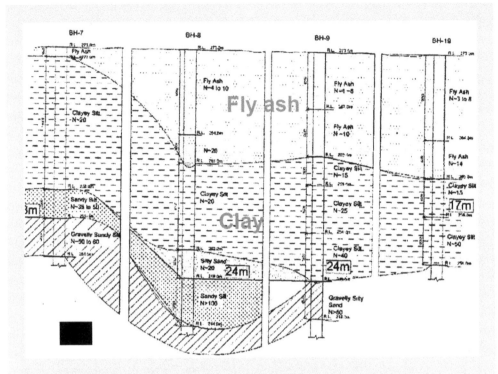

*Figure 3.30* A typical sectional profile showing soil conditions

Vibrofloat (vibrating unit)
Length = 2–3 m
Diameter = 0.3–0.5 m
Mass = 20 kN (lowered into the ground and vibrated).

A typical photograph of a vibrofloat is shown in Figure 3.31.
Why vibro stone columns method?

Improve bearing capacity of open foundations (100 kN/m²)
Enhance lateral capacity of piles (70 kN)
Mitigate liquefaction potential
Improve bearing capacity of open foundation: Vibro stone column of diameter 0.9 m
    at 2 m center-to-center spacing in a triangular grid pattern resulted in a bearing
    capacity of 100 kN/m².

Test after installation of stone column: A plate load test was conducted to access the
outcome, and the result at two locations is shown in Figure 3.32.
Improving lateral load capacity of piles: Vibro stone columns were installed in a specified
pattern surrounding the bored cast in situ (BCIS) piles to enhance the density of fly ash
deposits, which in turn can improve the lateral load carrying capacity (Figure 3.33).

*Figure 3.31* A photograph of vibrofloat

Source: http://powerlift.in/vibro-float

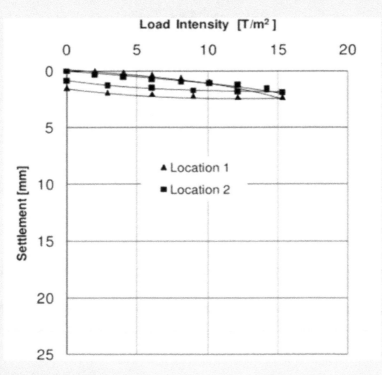

*Figure 3.32* Results of pile load test

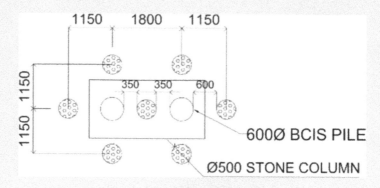

*Figure 3.33* Details of stone column surrounding a BCIS pile

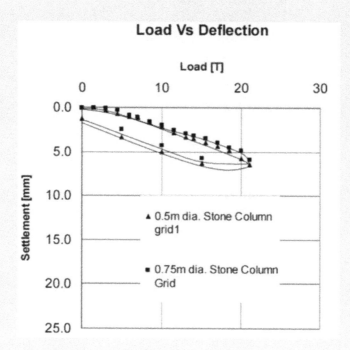

*Figure 3.34* Lateral test result

Tests were performed to access the lateral capacity, and a typical result of the lateral test is shown in Figure 3.34.

After improvement, result reported:

Design lateral load capacity = 70 kN
Ultimate load = 200 kN.

Source: Technical Report on Ground Improvement and Geosynthetics, IGS, New Delhi

## Case study 4

### *Shuaiba IWPP III – Desalination Plant – Saudi Arabia*

The Shuaiba Independent Water and Power Project (IWPP) was planned to meet the growing demands of water and electricity in Saudi Arabia's Shuaiba region, 110 km from Jeddah.

The site had two types of soil profiles. In the first profile there was loose to dense silty sand. In the second profile there was soft silt or very loose silty sand. This layer was followed by the bedrock.

The project consisted of 12 evaporators, three water tanks and a number of related buildings. The tank's diameter and height were 106.6 m and 20 m, respectively. The design criteria stipulated a bearing capacity of 200 kPa and maximum settlement of 75 mm for the tanks. For the other structures, the same were required to be 150 kPa and 25 mm, respectively.

Due to the presence of loose sands and soft silts, it was decided to optimise the foundation solution by implementing dynamic compaction and dynamic replacement in the project. The choice of this technique was dependent on the soil characteristics.

Upon completion of soil improvement works, 75 pressure meter tests (PMT) and one zone load test were used to demonstrate that the acceptance criteria had been achieved. The results of the tests clearly indicated the success of the ground improvement project and the ability of the foundations to safely support the design loads. Typical images from the site are shown in Figure 3.35.

*Figure 3.35* Typical photographs of during dynamic compaction

Source: Technical Report on Ground Improvement and Geosynthetics, IGS, New Delhi, 2016

## Case study 5 Dynamic compaction for T.C.L. fertilizer complex at Babrala, U.P. (India)

- The soil at Babrala consisted of a surface layer of loose, silty, sandy clay of 1–2 m depth underlain by loose fines and depths of 10–12 m. This in turn is underlain by silty, sandy clay.
- Parameters available at the site before the treatment indicate that the allowable net bearing capacity was 60 kPa. A seismic risk analysis of the site fixed the design earthquake as one of magnitude 6.4 with a peak acceleration of 0.2 g, which could induce significant liquefaction. A detailed testing program was made and the demarcation was done as shown in Figure 3.36.

The effectiveness of this technique at the site was established by treating two areas, each 30 m², by dynamic compaction.

Measurement of improvement was done by SPT and SCPT testing. The results of the exercise are shown in Figure 3.37. The results also demonstrate the increase in strength with time.

Targeted response and treatment:

- Based on the results from trials, modifications were introduced to obtain an allowable bearing pressure of 200 kPa at 2 m depth and that no liquefaction will occur in the improved ground during the design earthquake.
- The treatment consisted of four passes. The first pass was with a 10-ton hammer falling 16 m. The second pass was similar, but the locations are staggered. The third pass was with a 15-ton hammer falling 16 m. The final pass was with a 5-ton hammer falling 16 m on a grid of 2.5 m².

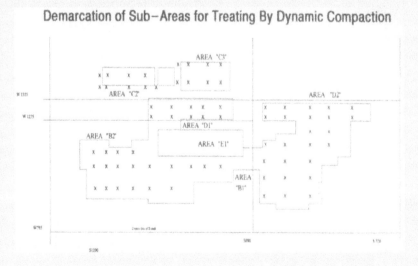

Demarcation of Sub–Areas for Treating By Dynamic Compaction

*Figure 3.36* Demarcation of testing site

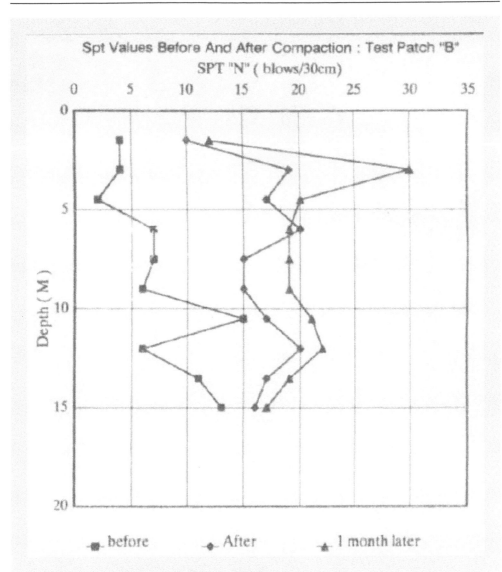

*Figure 3.37* SPT values before and after compaction

Quality monitoring:

- The treated soil was by the SPT N values as assurance against liquefaction and allowable bearing pressure are specified in terms of SPT N values obtained.
- The area treated was divided into sub areas as shown earlier and the results of the program are shown in Figure 3.38a–c.

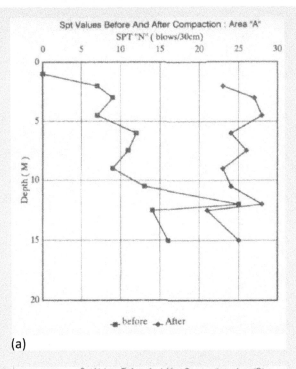

(a)

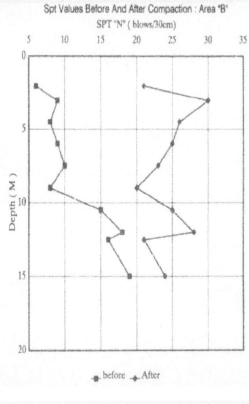

(b)

Figure 3.38 SPT test results

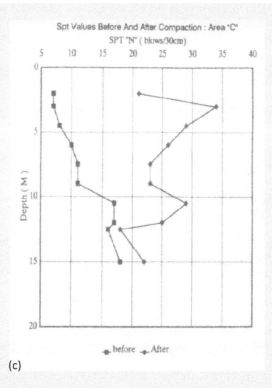

Figure 3.38  (Continued)

Conclusion:

• Dynamic compaction was successful in significantly increasing the strength of the soil. This translates to a more than threefold increase in bearing capacity over that of the initial design recommendation prior to treatment.

• The soils treated were loose sands to a depth of 12.5 m. Bearing capacities were increased from 60 to 200 kPa and the site was earthquake-proofed to the design earthquake.

Source: Sivakumar, 2018, 2016; www.nptel.ac.in

## References

Al-Refeai, T. O. (1992) Strengthening of soft soil by fiber reinforced sand column. *Proceedings of International Symposium of Earth Reinforcement Practice*. Fukuoka, Kyushu, Japan, pp. 11–13.

Arulrajah, A. & Abdullah, A. (2002) Vibro replacement design of high speed railway embankment. In: Huat et al. (eds) *2nd World Engineering Congress – Geotechnical Engineering and Transportation*. Universiti Putra Malaysia Press, Kuching, Sarawak, Malaysia. pp. 157–164.

Bell, F. G. (1975) *Methods of Treatment of Unstable Ground*. Newnes-Butterworth, London.

Biringen, E. & Edil, T. B. (2003) Improvement of soft soil by radial preloading. In: Huat et al. (eds) *Proceedings of 2nd International Conference on Advances in Soft Soil Engineering and Technology*. Universiti Putra Malaysia Press, Putrajaya, Malaysia. pp. 741–752.

Das, A. K. & Deb, K. (2017) Response of cylindrical storage tank foundation resting on tensionless stone column-improved soil. *International Journal of Geomechanics, ASCE*, 17(1), 04016035.

Edil, T. B. (2003) Recent advances in geotechnical characterization and construction over peat and organic soils. In: Huat et al. (eds) *Proceedings of 2nd International Conference on Advances in Soft Soil Engineering and Technology*. Universiti Putra Malaysia Press, Putrajaya, Malaysia. pp. 3–25.

First Report of the Technical Committee on Ground Improvement and Geosynthetics Indian Geotechnical Society, New Delhi, 2016, pp. 446.

Huat, B.B.K. & Ali, F. H. (1992) Embankment on stone columns: Comparison between model and field performance. In: *Proceedings Geotech. '92. Prediction Versus Performance in Geotechnical Engineering, Bangkok, 30 November–4 December*. pp. 373–385. Republished by A. A. Balkema, Rotterdam. (1994) Editors A. S. Balasubramaniam et al. pp. 167–174.

Huat, B.B.K., Craig, W. H. & Ali, F. H. (1993a) Behavior of sand columns improved foundation. *Journal Institution of Engineers, Malaysia*, 53(1), 65–74.

Huat, B.B.K., Craig, W. H. & Ali, F. H. (1993b) Reinforcing of soft soils with granular columns. *Proceedings of Eleventh South East Asian Geotechnical Conference, May, Singapore*. Southeast Asian Geotechnical Society, Singapore. pp. 357–361.

Kwong, H. K., Lien, B. & Fox, N. S. (2002) Stabilizing landslides using rammed aggregate piers. *Fifth Malaysian Road Conference, Kuala Lumpur, 7–9 October 2002, Malaysia*. Kuala Lumpur.

Lukas, R. G. (1995). Geotechnical Circular No. 1 – DYNAMIC COMPACTION, Federal Highway Administration Report FHWA-SA-95-037, March.

Maddison, J. D., Jones, D. B., Bell, A. L. & Jenner, C. G. (1996) Design and performance of an embankment supported using low strength geogrids and vibro compacted columns. *Geosynthetic: Application, Design and Construction*. Balkema, Rotterdam, The Netherlands. pp. 325–332.

Menard, L. & Broise, Y. (1975) Theoretical and practical aspects of dynamic consolidation. *Geotechnique*, 25, 3–18.

Priebe, H. J. (1995) The design of vibro replacement, *Journal of Ground Engineering*, 28(12), 31–37.

Sin, P. T. (2003) Economical solution for roadway embankment construction on soft compressible soil at Putrajaya, Selangor. In: Huat et al. (eds) *Proceedings of 2nd International Conference on Advances in Soft Soil Engineering and Technology, Putrajaya*. Putrajaya, Malaysia. pp. 649–657.

Sivakumar Babu, G. L. (2018) *Ground Improvement* (lecture 7). Available from: www. http://nptel.ac.in.

Sivakumar, Babu, G. L., Madhavi Latha, G., Madhav, M. R., Rajagopal, K., Mittal, S., Shah, D. L., Hari Krishna, Y. & Korulla, M. (2016) *Report of the Technical Committee on Ground Improvement and Geosynthetics*. Indian Geotechnical Society, New Delhi. pp. 446.

Sulaeman, A. (2003) An alternative foundation system on very soft soil. In: Huat et al. (eds) *Proceedings of 2nd International Conference on Advances in Soft Soil Engineering and Technology*. Universiti Putra Malaysia Press, Putrajaya, Malaysia. pp. 617–623.

Wissmann, K. J., Fitzpatrick, B. T., White, D. J., and Lien, B. H. (2002). Improving global stability and controlling settlement with Geopier soil reinforcing elements. *Proceedings, 4th International Conference on Ground Improvement*. Kuala Lumpur, Malaysia, 26–28 March.

# Chapter 4

# Replacement method, stage construction, preloading and drainage

## 4.1 Replacement method (excavation and backfilling)

One solution to treat soft compressible ground is to replace the poor soil by excavation or by dumping suitable imported fill materials if the subsoil is of a very high liquid type, as illustrated in Figure 4.1. This is naturally very expensive regarding materials. Also, it is difficult to control the underground movement of the material. In addition, there must be an environmentally acceptable location to waste the excavated soil within an economically acceptable haul distance and there must be a source of adequate fill again within an economically acceptable haul distance (Jarrett, 1997).

Soil replacement can be carried out when the groundwater surface is below the depth of excavation. If the groundwater surface is above the planned depth of excavation, replacement can be performed by the dredge method. The dredge is a method where excavation is made without pumping water out from the trench. After excavation is performed, the trench is filled with soil by a bulldozer. In the end, the new stratum of strong soil is compacted.

### 4.1.1 Design

For cases of excavation, the amount of soil that needs to be excavated ranges from full excavation to partial excavation of the compressible soil layer. The remaining layer will consolidate with time due to the reduction in the layer thickness.

The quantity of fill required and cross-section of the excavation is determined by embankment geometry and the engineering properties of both the subsoil and fill. Other factors that influence the excavation cross-section design are groundwater level and duration of the construction. It must be remembered that if excavation were made below the water table, then the fill would be difficult to compact. In this case consideration must be made for using materials that do not need compaction. The degree of compaction depends upon the type of fill used, as shown in Figure 4.2.

There are a number of empirical methods that could be used to determine the cross-section of an excavation. Figure 4.3 shows an example for a road embankment.

For organic soil, the width of the cross-section can be determined by taking into account the influence of vertical and horizontal deformation that will occur in the soil adjacent to the embankment foundation. The danger of damage can be reduced, and using loading berms constructed out of the excavated material itself can minimise the size of the excavation.

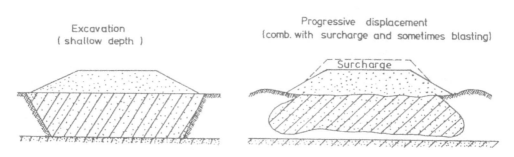

Figure 4.1 Excavation and replacement

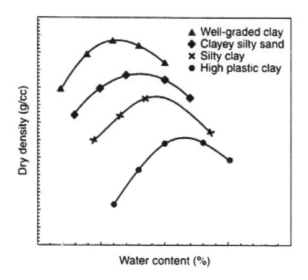

Figure 4.2 Compaction curves for different type of soils

Source: www.soilmanagementindia.com

### 4.1.2 Construction technique

Various equipment can be used to excavate soil above the water table. But for damp conditions below the water table, a dragline or an excavator is the most suitable.

Tight control during the excavation process is needed to ensure pockets of soft materials are not ignored. This is because this may lead to problem of differential settlement, lateral displacement and instability. The danger can be minimised by carrying out a continuous operation of excavation followed by replacement with suitable fill materials.

Ali and Abdul Wahab (2003) describe the application of the replacement method for construction of temporary rock bunds on a seabed for a land reclamation project.

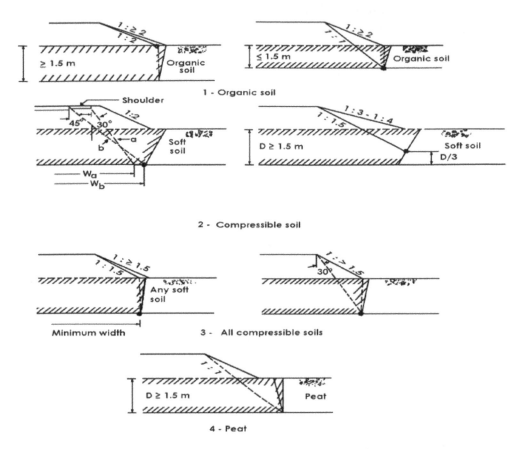

*Figure 4.3* Empirical method to determine width of excavation

## 4.2    Stage construction

Increase in effective stress of the subsoil will increase the shear strength of the subsoil. Therefore a problem with soil with initial strength which may be too low to carry, say, the full height of an embankment can be overcome by constructing the embankment in stages, to allow for sufficient gain in shear strength of the subsoil at each of the stages. Construction is carried out with adequate rest time between each loading increment (embankment height) in order to ensure that excess pore water pressure generated in the subsoil by the surface loading dissipates. The increase in soil strength results from the process of consolidation (increase in effective stress) to enable the subsequent load increment to be supported. This is the fundamental principle of stage construction.

This method is used in cases where the permeability of the subsoil is sufficiently large to allow adequate consolidation and hence strength, or in cases where the thickness of the underlying stratum is sufficiently thin to enables the excess pore water pressure to dissipate within the available time of construction. Suitable soils include silt, silty clay, clay with silt and sand layers, fibrous peat and other soils with permeability in the range of $10^{-6}$ to $10^{-7}$ cm/s.

In situation where these conditions are not satisfied, this method could also be used but in association with other techniques, such as vertical drains.

With very soft soil, the required shear strength of the soil may be safely maintained below the initial undrained shear strength of the soil using embankment slope reduction or the loading berms method. Reducing the embankment slope or incorporating a loading berm will result in a deeper critical slip surface and change the shape of the slip surface to a longer form. Due to the influence of this technique on the depth of the slip surface, it is more effective in cases where the strength of the soil increases with depth. But this method will require a larger quantity of fill material compared with the construction of a normal embankment.

### 4.2.1   Design and construction

In general practice, an embankment is first constructed to a height that corresponds with the shear strength of the in situ soil (usually with a factor of safety against slope instability of 1.2–2.0). Subsequent construction, with regards to rate and quantity of the construction, is generally based on continuous monitoring of pore water pressure and subsoil deformation.

A thorough site investigation has to be done. This is because the construction program is dependent on the increase in the shear strength of the soil. The stability of the embankment during construction should be evaluated. For this, the following methods can be utilised:

- Slices (Bishop, Morgenstern Price)
- Total stress analysis in clays (Shansep).

The undrained shear strength: $S_u = f(p')$. Also, $S_u/p' = f$(Liquidity Index) $= 0.1$–$0.3$
It should be noted that $p' = \sigma'_{0v}$ for normally consolidated clays.
The embankment can fail in shear, as shown in Figure 4.4.

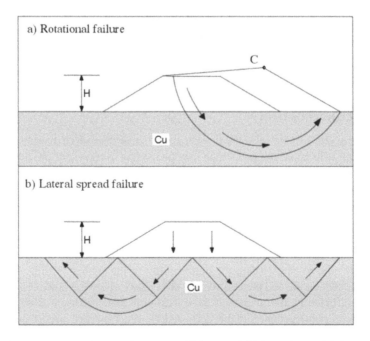

*Figure 4.4* Embankment shear failure: (a) rotational failure, and (b) lateral spread failure

Source: www.civil.uwaterloo.ca

The factor of safety of the embankment can be worked out using Equation (4.1).

$$F = \frac{N_c C_u}{\gamma_e H_e}$$ (4.1)

where

$F$ = factor of safety (typically 1.2)
$N_c$ = bearing capacity factor (to be calculated from Equation (4.2))
$C_u$ = undrained strength of soil
$\gamma_e$ = embankment unit weight
$H_e$ = height of embankment.

In the design, $C_u$ is the average undrained strength to a depth approximately equal to the height of the embankment. Since soil will fail at the weakest point, one should use average $C_u$ values in the soft zone.

$N_c$ can be calculated from Equation (4.2). It usually varies from 5 to 8.

$$N_c = 5 + 4\left(\frac{d_1}{d_2} - 0.4\right)$$ (4.2)

The steps to be followed in staged construction can be summarised as:

1. Construct embankment to $H_1$ (Figure 4.5)
2. Allow 90% of the excess pore water pressure to dissipate

   a. Estimate using consolidation theory
   b. Monitor piezometer installed in the soft zone

3. Determine increase in $C_u$ due to increase in vertical effective stress
4. Determine the magnitude of settlement
5. Increase height to $H_2$
6. Repeat steps 2 to 5 to get to design height.

The embankment height at any time can be calculated by Equation (4.3).

$$H = \frac{N_c}{F}\left(\frac{C_u}{\sigma_v'}\right)\left(\frac{\sigma_0'}{\gamma} + UH_1\right)$$ (4.3)

where $U$ is the average degree of consolidation over the potential failure zone.

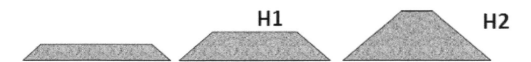

H1　　　　　　　　H2

Figure 4.5 Steps in staged construction

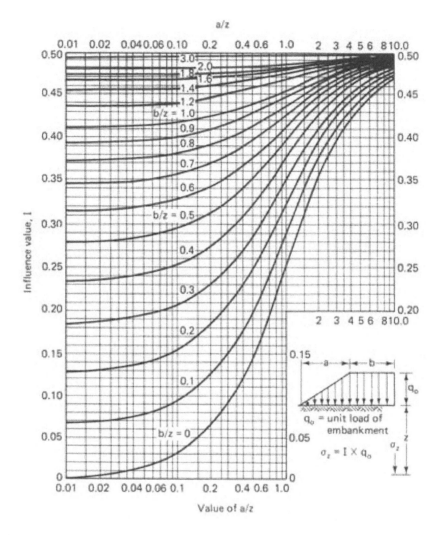

*Figure 4.6* Influence values for vertical stress under axis of a very long embankment

Source: After Osterberg (1957)

The magnitude of stress along the axis of the embankment can be estimated using the Osterberg chart (Figure 4.6) and under the corner of the triangular loading using Figure 4.7.

Usually the construction program extends up to 2.5 years to ensure adequate time to complete construction and to allow for majority of the consolidation settlement to have occurred.

Meanwhile, in cases where loading berms are used, their design (dimension, height, width and cross-section of the berms) depends on the counterweight required, the shear strength and the depth of the subsoil.

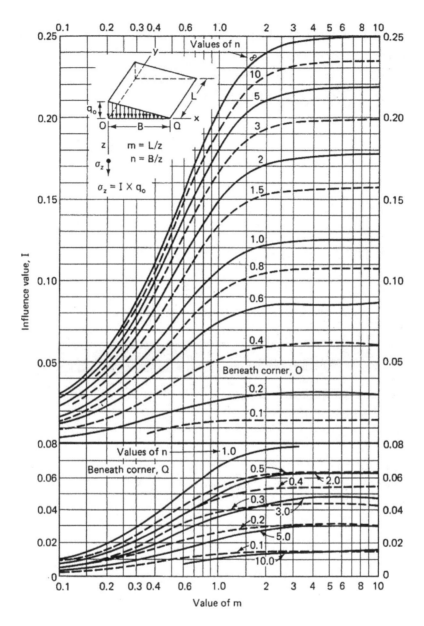

*Figure 4.7* Influence values for vertical stress under the corner of a triangular load of limited length

The usage of the berms, however, will increase the total settlement because of the increase in size of the loaded area. Furthermore, the settlement duration may be lengthened because of the increase in length of the horizontal drainage due to the berms. Figure 4.8 shows a schematic of loading berms.

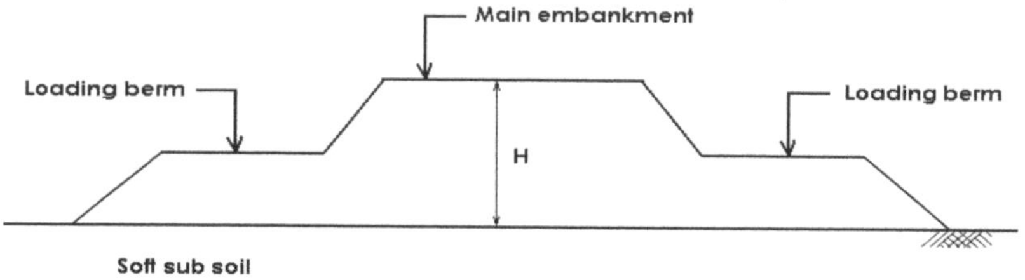

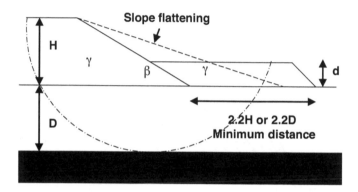

Figure 4.8 Schematic of loading berms

Figure 4.9 Effects of berms and slope angle

Source: www.civil.uwaterloo.ca

A schematic diagram of a berm is shown in Figure 4.9.

The use of berms can reduce the lateral spread of the embankment. This is more materially efficient than flattening the slope angle. The factor of safety can be calculated by Equation (4.4).

$$F = \frac{N_c C_u}{\gamma_e (H_e - d)} \tag{4.4}$$

where $d$ is the thickness of the berm. The $D_{max}$ must satisfy Equation (4.5).

$$F = \frac{N_c C_u}{\gamma_e d} \tag{4.5}$$

## 4.3  Preloading

The technique of preloading was developed in the 1940s and was used mainly in projects connected to highways. But nowadays this technique is found very useful for a variety of projects such as buildings, storage tanks, flood control structures, airfields and so forth. The drawback of this technique is that it does not provide favorable results for structures having heavy, concentrated loads. But this treatment is very effective for soils having high moisture content, high compressibility and low shear strength.

The magnitude of the preload pressure usually ranges from 1.2 to 1.3 times the actual structural pressure or is slightly greater than the maximum pressure that is generated by the proposed structural load. Earth fills (most popular), water lowering and vacuum under impervious membrane are some of the techniques used for applying preload.

The basis of preloading is to place a temporary fill over the construction site that is thicker than the final design fill. This causes settlement to occur more rapidly than would have occurred under the final fill design height. The preload is ideally left in place until it has settled more than the total amount that the design fill is expected to settle in its design life. Then the thickness of the preload fill is reduced to the final design thickness with the expectation that most of the settlement has now finished.

It is of interest to note that the method of preloading cannot be said to be a recent development, but it does still represent the method of construction with the best economic and engineering returns (Carlsten, 1988). It is however underused because of the necessity of prior planning.

They are two methods that could be used to apply the preload. They are by surcharging or by application of vacuum preload.

### 4.3.1  Surcharge

Preloading by surcharging involves constructing an embankment 10% to 20% higher than the final design. When the ultimate settlement under the design height has been reached, the surcharge will be removed. The use of preloading by surcharge is arguably the simplest and cheapest method to accelerate the process of consolidation. Therefore the method is usually used although its application is rather limited by the following factors:

• If a very long time is required to consolidate the subsoil, surcharge may require a very long time to achieve the desired degree of consolidation. This may not be suitable for certain construction.
• Shear strength of a number of soils is not sufficient to support the surcharge. Therefore excessive plastic deformation or even failure may occur.
• In certain cases, earth fill to be used as the surcharge may have to be imported from far away, therefore its cost may be high.

The preloading technique can also be combined with other methods, such as the vertical drain method. Some successful case histories have been described by Ali and Huat (1992a), and Han and Eng (2003).

Figure 4.10 shows a schematic of the preloading with the surcharge method.

### 4.3.2  Design and construction

The basic theory used in the design of a preload is based on the assumption that increase in effective stress at any given time is directly proportional to the magnitude of the surcharge load increment. The total settlement is therefore controlled by the magnitude of the effective stress. Increase in the surcharge load will therefore increase the settlement accordingly.

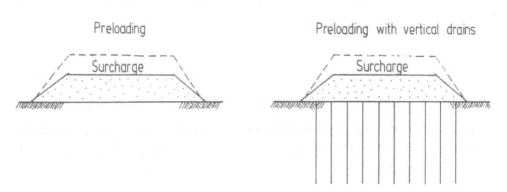

Figure 4.10 Preloading and vertical drains

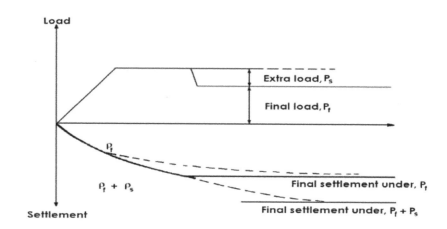

Figure 4.11 Principle of preloading

Referring to Figure 4.11, under final embankment load, $P_f$, the relationship between settlement $r_f$ and time $t$ can be obtained. The surcharge is removed after settlement $r_f$ has been obtained.

The magnitude of the surcharge ($P_s$) and its application duration is determined using the normal settlement calculation. It must also be remembered, however, that the surcharge should not be left longer than required. This is because if the surcharge is left too long, when it is later removed, the subsoil may experience swelling, resulting in damage similar to that caused by settlement.

### 4.3.3  Vacuum preloading

The technique of preloading with vacuum, as described by Kjellman (1952), is an alternative procedure of applying surcharge to the subsoil. The technique involves placing a layer of impermeable plastic membrane over an area to be treated, as shown in Figure 4.12. Vertical drains accelerate consolidation by providing short, horizontal drainage paths and are employed worldwide in many soft soil improvement projects. In this method the pore water pressure is reduced to below atmospheric level by a pumping system. Surcharge is applied through the difference between the atmospheric pressure which acts on the membrane and the negative pore water pressure.

This technique can be used to limit settlement and increase the shear strength of the subsoil. Vacuum preloading is often used in conjunction with vertical drains. This technique has one major advantage over the normal preload: no embankment construction is required. As such, the problem of soil instability will not arise. It enables the equivalent construction of a very high embankment on very soft ground to be made over a relatively short period of time by reducing the development of shear strain in the soil (Mitachi et al., 2003).

Prefabricated vertical drains (PVDs) have become an economical and viable ground improvement option because of their rapid installation with simple field equipment. The application of a vacuum load in addition to surcharge fill can further accelerate the rate of settlement to obtain the desired settlement without increasing the excess pore pressure. PVDs distribute the vacuum pressure to deep layers of subsoil, thereby reducing the excess pore water pressure due to surcharge. The consolidation process of vacuum preloading compared to conventional preloading is shown in Figure 4.13.

Indraratna et al. (2005b) introduced the unit cell analysis for vacuum preloading under instantaneous loading. However, while an embankment is being constructed on soft clay, the fill surcharge is usually raised over time to attain the desired height. Therefore, a time-dependent loading due to filling would be more appropriate than an instantaneous loading, especially during the initial stages of construction. Hence, the embankment load from filling ($\sigma_t$) is assumed to increase linearly up to a maximum value ($\sigma_1$) at time $t_0$ and kept constant

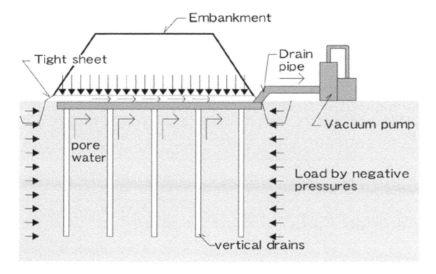

*Figure 4.12* Typical setup of vacuum preloading method

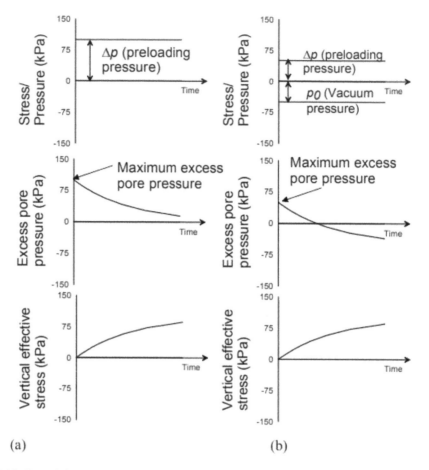

*Figure 4.13* Consolidation process: (a) conventional loading, (b) idealised vacuum preloading

Source: Modified from Indraratna et al. (2005a)

thereafter (Figure 4.14a). The vacuum is applied at $t = t_{vac}$. Figure 4.14b illustrates the unit cell adopted for analytical solutions with boundary conditions (Figure 4.14c).

Indraratna et al. (2011) proposed that the average excess pore pressure due to radial consolidation while considering the smear effect under time dependent surcharge $(\overline{u}_L)$ can be expressed by:

$$\overline{u}_L = \frac{\mu d_e^2}{8 c_h t_0}\left(1 - exp\left(\frac{-8 c_h t}{\mu d_e^2}\right)\right)\sigma_1 \quad for\, 0 \le t \le t_0 \tag{4.6}$$

$$\overline{u}_L = \frac{\mu d_e^2}{8 c_h t_0}\left(1 - exp\left(\frac{-8 c_h t}{\mu d_e^2}\right)exp\left(\frac{-8 c_h (t - t_0)}{\mu d_e^2}\right)\right)\sigma_1 \quad for\, t > t_0 \tag{4.7}$$

where $d_e$ = influence zone diameter, $c_h$ = coefficient of consolidation for horizontal drainage, $\sigma_1$ = applied surcharge pressure and $t$ = time.

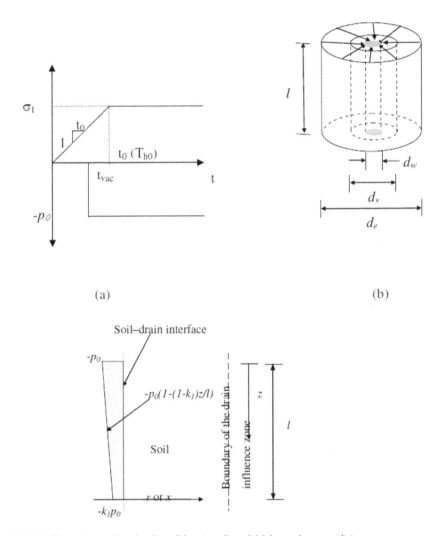

*Figure 4.14* (a) Time-dependent loading, (b) unit cell, and (c) boundary conditions

Source: Rujikiatkamjorn and Indraratna (2013)

Recently, Indraratna et al. (2005b) showed that the average excess pore pressure under radial consolidation due to vacuum pressure ($u_{vac}$) alone could be determined from:

$$u_{vac} = 0 \qquad t < t_{vac} \tag{4.9}$$

$$u_{vac} = p_0 exp\left(\frac{-8c_h\left(t - t_{vac}\right)}{\mu d_e^2}\right) - p_0 \tag{4.10}$$

where $d_e$ = the diameter of the soil cylinder dewatered by a drain, $d_s$ = the diameter of the smear zone, $d_w$ = the equivalent diameter of the drain, $k_s$ = horizontal soil permeability in the smear zone and $\mu =$ a group of parameters representing the geometry of the vertical drain system and the smear effect. The $\mu$ parameter can be given by:

$$\mu = ln\left(\frac{n}{s}\right) - \frac{3}{4} + \frac{\kappa(s-1)^2}{(s^2 - 2\kappa s + \kappa)} ln\left(\frac{s}{\sqrt{\kappa}}\right) - \frac{s(s-1)\sqrt{\kappa(\kappa-1)}}{2(s^2 - 2\kappa s + \kappa)} ln\left(\frac{\sqrt{\kappa} + \sqrt{\kappa - 1}}{\sqrt{\kappa} - \sqrt{\kappa - 1}}\right) \quad (4.11)$$

In the above expression, $\kappa = k_h/k_0$ and $k_0 =$ minimum permeability in the smear zone.

### 4.3.4   Design method

The preload intensity and duration of vacuum application are determined in the same manner as those described for the surcharge method.

### 4.3.5   Construction technique

The vacuum preloading operation involves first placement of a layer of sand about 150 mm thick over an area to be treated. The sand is then wrapped with an impermeable membrane with its corners tucked into shallow trenches to yield an effective vacuum (Figure 4.12). Vacuum with a high pressure (60–80 kPa) is then applied in the sand layer using a pumping system to generate an equivalent surcharge on the earth surface.

The method is particular suitable for treating very soft soil which has little capacity to support even a light load without excessive deformations, such as peat.

## 4.4   Drainage methods

Construction projects such as foundation for roads, railways, embankments and airport runways are examples of spread construction that involves large areas with variable soil properties. These areas may include soft compressible soils areas such as alluvium. Such constructions will not only give rise to the problem of instability during construction but also to problems with long-term and persistent settlement thereafter. One solution is to make use of the stage construction technique as described earlier. This solution, however, requires time for the excess pore water that has been generated to dissipate and for the corresponding shear strength of the soil to increase through consolidation. Depending on the permeability of the in situ soil, the stage construction method may need a construction time, which is much longer than permissible for the particular project. However this can be rectified by using the drainage method.

A drainage system is normally used to accelerate consolidation by facilitating discharge of excess pore water pressure by reducing the drainage path length. Both horizontal and vertical drainage systems are used. Of the two, the vertical drainage method is commonly used for treating thick deposits of soft soils.

The technique for installation of vertical drains was described in the early 1930s. Many systems of vertical drain exist nowadays including the vertical sand drain, cardboard wick drain and plastic strip drain. All these drains are installed in the same manner through a compressible soil layer in the form of a triangular or square grid in plan. The spacing between drain can be 1 m or more, depending on the drain type and design requirements.

Meanwhile, the horizontal drainage system is usually constructed in form of trenches dug into soft soil at a specific interval. These trenches are then backfilled with free draining granular materials. Sometimes perforated pipes are included to increase the effectiveness of the drain.

Another drainage technique is based on the principle of electro-osmosis. It has long been known that electric current can cause pore water to flow from a positive anode pole to a negative cathode pole. The application of this method (also sometimes known as electro-drainage) results in the following effects to soil:

- The soil moisture content is reduced and thereby results in soil with reduced compressibility and increase in shear strength (undrained cohesion). With this the soil settlement can be reduced, and the stability of an embankment constructed on the treated soil can be increased.
- The water content that increases in the soil surrounding the cathode will reduce friction, thus enabling the cathode to be pulled out easily.

The use of this method is rather limited, however.

### 4.4.1  Consolidation process

Construction of any embankment for a project like road construction, a railway or simply to increase platform level of a building project will exert pressure on the subsoil. For cases where the subsoil comprises sand or gravel, the settlement that results is usually small and will occur over a short period of time. But for cases where the subsoil consists of a layer of compressible soils like soft clays, peat or loam, the increase in stresses usually results in settlement of a large magnitude and requires a very long time to complete if the soil is not treated, and particularly if the soil layer is very thick.

Pore water pressure that results due to the surface loading on soil with low permeability, such as saturated clay, generally cannot be dissipated immediately. The pressure will therefore reduce the soil effective stress in a short term. Because of this, the factor of safety against soil instability also reduces in the short term. But with time, this excess pore water pressure will dissipate. The effective stress of the subsoil will increase, as will the shear strength of the soil, and hence there will be an increase in the factor of safety against instability. Therefore, an embankment built on saturated soft soil is more stable in the long term than the short term. To increase the rate of dissipation of excess pore water pressure, drainage methods as described above could be used.

Soil consolidation process and the distribution of vertical stress with depth can be described as follows. The soil is assumed to be fully saturated.

$$\sigma = \sigma' + u \tag{4.12}$$

where

$\sigma$ = total stress
$\sigma'$ = effective stress
$u$ = pore water pressure.

Equation (4.12) (Terzaghi, 1925) provides the relationship between effective stress, total stress and pore water pressure, which is considered true for a fully saturated soil. Laughton

(1955), in his experiment with lead balls inside an oedometer, proposed the following refinement to Terzaghi's original equation:

$$\sigma' = \sigma - (1 - c_s/c)u \tag{4.13}$$

where $c_s$ is the soil skeleton compressibility and $c$ is the compressibility for volume change. Skempton (1960) shows, however, that for most soil materials the ratio of $c_s/c$ is very small. In this case, Laughton's equation will become similar to Equation (4.12) as proposed by Terzaghi.

For example, a subsoil comprises a layer of saturated clay 10 m thick. The site is backfilled with sand 4 m high. This filling causes a surcharge load of 80 kN/m² onto the subsoil (that is for sand with unit weight $\gamma$ of 20 kN/m³). The total settlement in the subsoil under this new load is due to compression of the clay layer and elastic compression of the sand layer. In any case, the elastic compression of the sand layer is normally small and can be ignored.

As mentioned earlier, clay usually has low porosity and low permeability. Water inside the soil void cannot be compressed, as compared with the soil skeleton. The clay layer is also generally not permeable enough to allow free drainage of water immediately. Therefore the applied load will have to be first carried by the pore water itself, causing the pore pressure to increase.

This increase in pore pressure inside the void space is called the "excess pore water pressure." This excess pore water pressure will cause the water to flow through the soil void. In the example in Figure 5.8, water will flow to the sand layer above and below the clay layer. This is because sand has higher permeability compared with clay. The total water volume will be reduced. Soil voids will therefore become smaller when water under the excess pressure is gradually discharged. In this process, total soil pressure, $s = (s' + u)$, is assumed to remain constant. Therefore, the reduction in pore pressure is similar to the increase in effective stress of the soil. As soon as all this excess pore pressure has dissipated, a phenomenon called the consolidation is completed. In the consolidation process (actually it is a drainage process), soil excess pore water pressure will slowly dissipate with time. The soil permeability will also reduce with time. Because of these factors, the soil rate of settlement actually reduces with time. In theory, the process of consolidation can never be completed.

The consolidation period for a soil layer varies not only due to its permeability but also due to the layer thickness. For example, a 10 m thick clay layer may need 30 years to consolidate compared with say 270 years for a 30 m thick layer of the same permeability. Therefore it can be summarised that the duration of soil consolidation is mainly dependent on two factors:

• Permeability of the compressible soil layer
• Maximum distance that the water has to travel before achieving hydrostatic equilibrium. This distance may be a half-layer thickness in cases where the upper and lower soil layers bordering the compressible layer are sufficiently permeable, or it can be equal to the entire compressible layer thickness if the water can be discharged in one direction only – for example, a clay layer overlying solid (relatively impermeable) bedrock.

The numerical analysis for the consolidation process is complicated because of the following factors:

- The permeability of the compressible soil may vary a lot over depth and in different directions. Ordinary laboratory tests, such as an oedometer test on a small size sample, is normally unable to show this difference. This has been proven by Rowe (1972).
- We do not actually have accurate information on the process of how groundwater flows under excess pressure.
- The direction of flow is not only limited to vertical direction as normally assumed in simple analysis. This may only happen in case of a very homogeneous soil with an infinitely large applied load. In reality, real soil is seldom homogeneous.
- Soil voids are not necessarily all filled with water. The voids may also contain air. For partly saturated soil, Bishop (1959) proposed the following equation for effective stress:

$$\sigma' = \sigma - u_a + \chi(u_a - u_w) \tag{4.14}$$

where $u_a$ is the air pore pressure, $u_w$ is the water void pressure and $\chi$ is the coefficient of saturation. For a fully saturated clay, $\chi = 1$. Equation (4.14) will then become similar to Terzaghi's (1925) equation, that is:

$$\sigma' = \sigma - u_w$$

Jennings and Burland (1962) have shown that for partly saturated soil below the critical degree of saturation, the behaviour of the soil could not be accounted for based on the principle of effective stress only. In the case of sand, this critical degree of saturation is below 50%. However, for clay the critical degree of saturation can be as high as 90%.

- The total stress during the process of consolidation actually reduces because the thickness of layer above the water table is reduced.

Because the magnitude and rate of settlement are closely related to the thickness of the compressible layer and settlement rate parameter, $c_v$, we need to perform a thorough site investigation. Any structure or fabric, such as strips or layers of slit and sand in the clay layer, will significantly influence the rate of soil settlement.

### 4.4.2   Vertical drainage method

As mentioned earlier, there are three drainage methods that could be used to treat soft compressible soils. They are vertical drains, horizontal drains and electro-osmosis. Of these three methods, the vertical drain method is commonly used. Horizontal drains and electro-osmosis have limited applications.

As mentioned before, spread construction on soft compressible soil will give rise to a short-term stability problem and a long-term settlement problem. Therefore the subsoil needs to be treated. With the drainage method, placing drains inside the compressible subsoil shortens the drainage path of the water. This is shown in Figure 4.15. In Figure 4.15a, no drains are installed, and the water has to flow in one direction only. In Figure 4.15b, drains

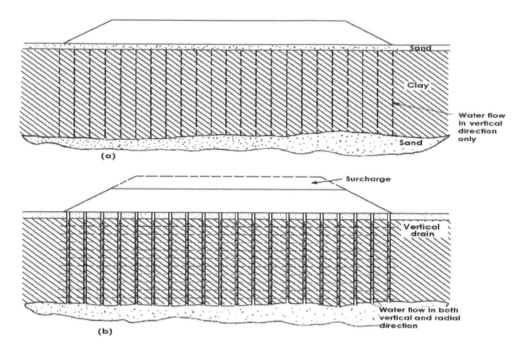

*Figure 4.15* (a) Without drain, (b) with vertical drain

have been installed, and water can now flow not only in the vertical direction but also in the horizontal (radial) direction to the nearby vertical drains. Therefore, in theory, the process of water draining under the excess pore pressure will be considerably accelerated with the aid of the vertical drains.

O. J. Potter in the 1930s introduced a method to accelerate consolidation with vertical drains consisting of sand drains. This method is also known as the "sand column" or "sand pile." The distance of center to center of the drain is about 2–3 m for drains of diameter 200–400 mm. These drains can be installed to depths of 30 m or more.

Parallel to the development of the sand drain, the strip (or wick) cardboard drain was invented by Kjellman in Sweden. The strip drains are about 4 mm thick by 100 mm wide with an open central core. These drains were widely used for construction of roads and runways in Sweden. The cross-sectional area of the cardboard drain is much smaller compared with the sand drain. Therefore the cardboard drains need to be placed much closer together (e.g., 0.8–1.2 m center to center in plan compared with 2–3 m for sand drain) than the sand drains. The cardboard drains are installed into the ground using a steel tube (also known as a mandrel), which is inserted into the ground.

Besides sand drains and Kjellman's cardboard drains, which can be considered traditional vertical drains, nowadays there is a new generation of vertical drains, that is the prefabricated drain in strip form. Most of these prefabricated drains are made of synthetic (plastic) materials. Though in general these drains function in the same manner as the cardboard drains, they are different in terms of their compositional material. These drains are usually made of a central core wrapped with a thin jacket layer made of woven fabric or paper. The

*Figure 4.16* Prefabricated vertical drain

Source: www.indiamart.com

main function of the central core is to facilitate free drainage of water, while the jacket (fabric or paper) functions to prevent soil fines from getting into the central core, causing it to clog. The process of water draining into a vertical drain is shown schematically in Figure 4.16.

Ideally, the filter jacket must initially be permeable enough to allow fine soil particles to pass through, thereby forming a bridging network of larger particles adjacent to the drain. This will form a natural grade filter in the soil. But it must be remembered that if the pores of the fabric jacket are too large, continuous piping will occur. This will then cause the drain to clog, thus making it ineffective. A relatively stiff woven geotextile filter jacket seems to be the most favorable (Ali and Huat, 1992b).

Nowadays there are many types of prefabricated drains available on the market. Some of these are as follows:

1.  Colbond drain (Figure 4.17(a)): This drain was invented by Akzo Research Laboratory in Holland. It comprises an Enkamat central core, which is made from coarse monofilament polyester with a labyrinth (three-dimensional) structure. The filter jacket is made of woven polyester. These drains are usually 4 mm thick and 100 mm wide, but larger widths of 150 mm and 300 mm are also available in the market.
2.  Geodrain: O. Wager at the Swedish Geotechnical Institute invented this drain. The drain is made of a grooved central plastic core wrapped with a filter jacket that is resistant to wet soil. The usual drain dimension is 4 mm thick and 100 mm wide.
3.  Membra drain (Figure 4.17(b)): This drain is manufactured by Geotechnic Holland B.V. As with the Geodrain, it comprises a grooved central core made of polypropylene surrounded by a filter jacket of woven polypropylene or paper.

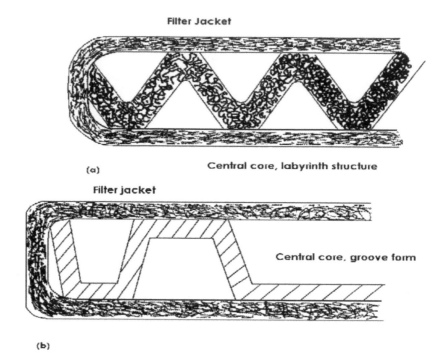

Figure 4.17 Cross-section: (a) Colbond drain, (b) Membra drain

### 4.4.3   Design of vertical drainage system

As mentioned earlier, the main objective of using a drainage method is to accelerate the rate of consolidation settlement. In the case of a vertical drain, the target degree of consolidation settlement at the end of construction is between 80% and 90%. In this case, the shear strength as well as the stability of the subsoil will be improved, while settlement – in particular differential settlement – will be reduced. This can be achieved by reducing the length of drainage that water needs to go through under excess pressure. Therefore the spacing between center to center of the drain is an important design consideration.

The drains are usually installed in a triangular or square grid in plan, as shown in Figure 4.18. The square pattern will give equivalent radius R = 0.564 S, while rectangular pattern gives equivalent radius R = 0.525 S (where S is the center to center spacing of the drain). According to Elzen and Atkinson (1980), the rectangular pattern is closest to a cylindrical model assumed for the drain design, compared with the square pattern.

It is important in design to know the values of soil coefficient of consolidation in both horizontal and vertical directions ($c_h$ and $c_v$). The ratio of $c_h/c_v$ is normally in the range of 1–2. The higher the ratio, the more effective would be the installation of the vertical drain in accelerating soil drainage, hence consolidation. Besides this, when designing for vertical drains, soil geological features such as thin sand or silt layers need to be taken into account. It must also be remembered that the permeability of soil adjacent to the drain may be reduced due to the remolding effect during the process of drain installation, particularly if a mandrel is used. This effect is known as smear. The smear effect can be accounted for in the design

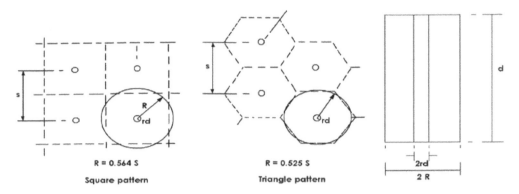

*Figure 4.18* Influence zone of vertical drain

by reducing the value of $c_h$ or by reducing the drain diameter (i.e., its circumference area). The remolding effect can also be reduced by exercising great care during installation of the drain, in addition to choosing a drain with a suitable filter jacket. However if the in situ soil is too sensitive, it is unlikely that the drain installed in it will be effective.

As mentioned earlier, the new generation of drains is in the form of strips. Their equivalent diameter can be obtained from the following:

$$d = 2(b + t)/\pi \tag{4.15}$$

where *d* is the equivalent diameter of an equivalent cylinder, *b* is the width of the drain and *t* is its thickness.

But Equation (4.15) does not take into account the edge effect on water flow, whereby water will concentrate to the edge of the drain. Van den Elzen and Atkinson (1980) have proposed a factor of $\pi/4$ to be used to reduce the equivalent diameter of the cylinder given in Equation (4.15). With this, the equation for calculating equivalent diameter of the cylinder becomes:

$$d = 2(b + t)/4 \tag{4.16}$$

As has been mentioned, the main function of a drain is to accelerate the process of consolidation of a loaded soil. In theory, there are three major components of soil settlement: (a) immediate settlement (elastic settlement), (b) consolidation settlement and (c) secondary settlement or creep. In designing with drains, it is normally the consolidation settlement that needs to be accelerated. Because of this, there is some controversy regarding the effectiveness of vertical drains in soil with high secondary settlement, such as high plastic clay or peat. Surficial peats often exhibit very high permeability until they are compressed, and a large part of their total compression takes place under constant effective stress. Vertical drains may be effective in accelerating strength gain (Kurihara et al., 1994) but not total settlement. The effectiveness of the drains may be additionally limited by deterioration and buckling of the drain and the consequent decline of discharge capacity. However, the general consensus is that vertical drains are effective tools for construction over peat (Edil, 1994).

In the previous sections we have seen the various types of drains, especially the prefabricated drains that are available in the market. In determining the most suitable drains for a particular job at hand, the following factors need to be taken into account:

1.  A large deformation in the subsoil may lead to damage and discontinuity of the drainage system
2.  The drain may clog due to piping of soil particles
3.  The materials use for making the drain, especially the prefabricated type, may rot.

Figure 4.19 shows a schematic of a drainage system used to accelerate consolidation using the vertical drain method, in addition to surcharge and loading berms. Surcharge, sometimes with loading berms, is used to reduce the number of drains required.

There are several methods that can be used to determine the dimension and spacing of the required vertical drains. The method based on radial consolidation theory by Barron (1948) is normally used.

A theoretical relationship between the diameter of a soil cylinder, $D$, with the degree of consolidation due to horizontal or radial drainage is given in Equation (4.17). The main assumption made is that the horizontal cross-section remains horizontal (equal strain theory). The relationship is given as:

$$t = \frac{D^2}{8C_h} F(n) \, In\left(\frac{1}{1-U_h}\right) \tag{4.17}$$

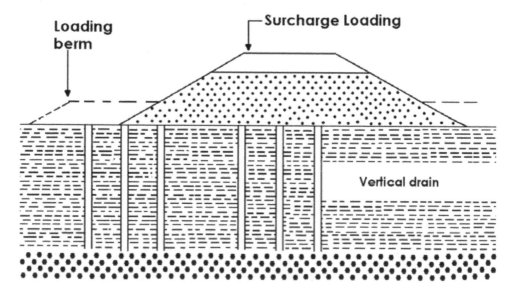

*Figure 4.19* Schematic of a drainage system used to accelerate consolidation using the vertical drain method

$F(n)$ = drain spacing factor = $In\left(\dfrac{D}{d_w}\right) - \dfrac{3}{4}$

where

$U_h$ = average degree of consolidation (for horizontal drainage only)
$c_h$ = coefficient of consolidation in horizontal direction (m²/yr)
$t$ = time required to achieve $U_h$ (years)
$D$ = diameter of soil cylinder (m) (i.e., diameter of influence zone)
$d_w$ = equivalent diameter of drain (m).

Subsequently Barron's solution can be combined with consolidation theory to give a numerical analysis that takes into account of dissipation of pore water pressure in both horizontal (radial) and vertical directions. This combined solution will be in three-dimensional form. It can be stated in polar coordinates, taking into account the variable soil properties as follows:

$$\frac{du}{dt} = c_h\left(\frac{d^2u}{d^2r} + \frac{1}{r}\frac{du}{dr}\right) + c_v\frac{d^2u}{dz^2} \qquad (4.18)$$

A vertical soil prism surrounding the drain (influence zone) is replaced with a cylindrical block with radius R and a similar cross-sectional area.

$$U_v = f\left(t_v\right) \qquad (4.19)$$

and

$$U_h = f\left(t_h\right) \qquad (4.20)$$

where $U_v$ is the average degree of consolidation due to vertical drainage; $U_h$ is the average degree of consolidation due to horizontal (radial) drainage; $t_v$ ($= c_v t/d^2$) and $t_h$ ($= c_h t/d^2$) are time factors for vertical and horizontal drainage, respectively; $c_v$ and $c_h$ are the coefficients of consolidation for flow in vertical and horizontal directions, respectively; and $t$ is the consolidation time.

The solution for radial drainage by Barron can be plotted in graphical form, as shown in Figure 4.20a. The relationship between $U_r$ and $t_h$ is dependent on the ratio $n(n = R/r_d)$, where R is the radius equivalent cylindrical block and $r_d$ is the radius of the drain. The solution for vertical drainage can be obtained using the one-dimensional consolidation theory. Figure 4.20b shows the relationship between degrees of consolidation (vertical) ($U_z$) with time factor $t_v$.

To calculate the average degree of consolidation, $U$, that is due to the combination of vertical and horizontal drainage, Equation (4.21) (Carillo, 1942) can be used:

$$\left(1 - U\right) = \left(1 - U_v\right)\left(1 - U_h\right) \qquad (4.21)$$

### 4.4.3.1    Example 1

An embankment is to be built on a clay layer 20 m thick, below which is a layer of permeable sand. Building the embankments adds an effective vertical stress on the order of 80 kN/m² to the clay layer. The following data are obtained from the laboratory (oedometer) test on the

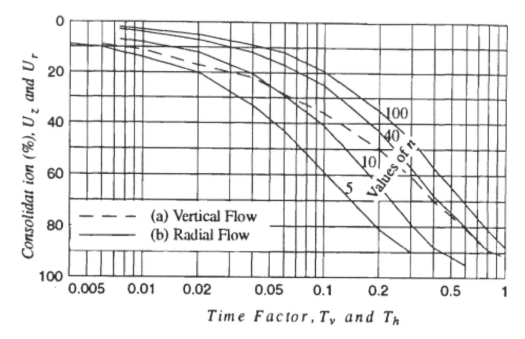

*Figure 4.20* Average consolidation rates: (a) for vertical flow, and (b) for radial flow

Source: After Barron (1948)

clay samples: $c_v = 5$ m²/yr, $c_h = 10$ m²/yr and $m_v = 0.2$ m²/MN. Allowable residual settlement after 6 months from end of construction is 30 mm. Assuming sand drains of diameter 400 mm are to be installed in a triangular grid spacing required to treat this site, determine the required drain spacing.

### 4.4.3.2  Solution

Total consolidation settlement $\left(1 \text{ dimension}\right)\rho \;=\; m_v\Delta\sigma H$
$$= 0.2 \times 80 \times 20$$
$$= 320 \text{ mm}$$

For $t = 6$ months, the average degree of consolidation that needs to be achieved is:

$$U = \frac{320 - 30}{320} = 0.91\%$$

Diameter of the sand drain is given as 400 mm. Therefore, $r_d = 200$ mm $= 0.2$ m
Radius of cylindrical block, $R = n.r_d = 0.2n$
Drainage path length, $d = H/2 = 20/2 = 10$ m (that is for vertical drainage in two directions)

$$T_v = c_v t/d^2 = 5 \times 0.5 / 10^2 = 0.025$$

For $T_v = 0.025$, $U_v = 0.21$

$T_h = c_h t/R^2 = 10 \times 0.5/4 \times 0.2^2\ n^2 = 31.25/n^2$

That is, $n = (31.25)/T_h$

$$\left(1-U\right) = \left(1-U_v\right)\left(1-U_h\right)$$
$$\left(1-0.91\right) = \left(1-0.91\right)\left(1-U_h\right)$$
$$0.09 \quad = 0.79\left(1-U_h\right)$$
$$U_h \quad = 0.89$$

Solution by trial and error:

| n | $T_h$ | $(31.25)/T_h$ |
|---|-------|---------------|
| 5 | 0.26 | 10.9 |
| 10 | 0.45 | 8.3 |
| 20 | 0.65 | 6.9 |

If we plot $n$ versus $(31.25)/T_h$, it is found that $n = 9$. Therefore R $= 0.2n = 0.2 \times 9 = 1.8$ m
The spacing of drain in a triangular pattern (see Figure 4.18) is:

S = R/0.564 = 1.8/0.564 = 3.2m

Beside the above solution, Barron's solution can also be simplified for value of $D/d > 6$ as follows:

$$t = \frac{D^2}{8c_h}(_{ln}\frac{(D)}{d} - \frac{3}{4})_{ln}\frac{1}{1-u_h} \tag{4.22}$$

For a prefabricated strip drain, the drain diameter can be calculated using Equation (4.16):

$$d = 2\left(b+t\right)/4$$

To simplify calculations, the value of width ($t$) of the drain is ignored. Equation (4.16) then becomes:

$$d = b/2 \tag{4.23}$$

If $U_v$ is ignored, and the drains are installed in triangular pattern in plan, then the graph that relates degree of consolidation (horizontal), time and coefficient of horizontal consolidation, $c_h$, can be derived. In this case, the width of the strip drain is 100 mm.

### 4.4.3.3 Example 2

A site of 2 ha is to be used for a housing project. The subsoil comprises 12 m thick soft clay with $c_h = 4$ m$^2$/yr. The site is to be backfilled with a 3 m thick sand layer. Determine the spacing and number of vertical drains required to achieve average consolidation, $U_h = 90\%$ after only 1 year.

#### 4.4.3.4   Solution

From the graph in Figure 4.21, it is found that for $c_h = 4$ m²/yr, and degree of consolidation $U_h = 90\%$ after 1 year, the required drain in triangular pattern is 2.04 m. The number of drains required is 5550.

It must be stated here that the preceding solution is only for homogeneous subsoil. In a more complex case, such as where the subsoil comprises several layers of compressible soils with varying properties, we may need to employ more complicated analysis such as the finite element or finite different methods. These methods, however, can only be employed if the project at hand is large enough to warrant comprehensive site investigation work to be carried out in order to obtain sufficient data for the design input.

### 4.4.4   Technique for installation of vertical drains

First a layer of granular material such as sand is placed on site as a working platform for the installation equipment. This layer of granular material also acts as a drainage layer to dissipate excess pore water pressure.

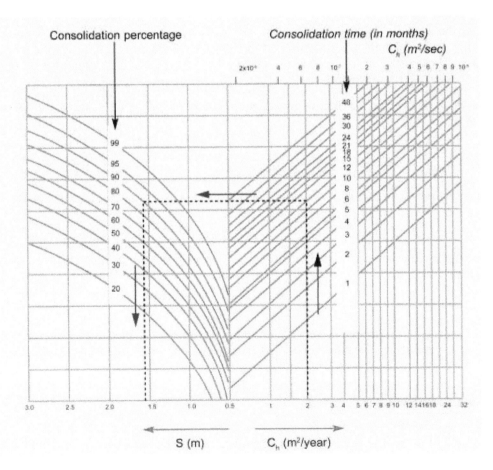

Figure 4.21 Vertical drain design curve

The installation technique to be used usually depends on the type of drains to be installed. Whichever system is used, it is very important to keep disturbances to the in situ soil to a minimum. This is because soil shear strength and its drainage characteristics are very much influenced by such disturbances.

For a sand drain, it is usually constructed by drilling a borehole into the ground either by using the water-jetting technique or by driving a mandrel. Sometimes casings are used to prevent collapse of the boreholes. A driven mandrel in general gives a higher degree of disturbance because of the large volume of soil that has to be displaced during the driving process. The jetting technique can reduce disturbances to the in situ soil. Once the hole has been made in the ground, it is then backfilled with suitably graded sand. The sand drains method, however, is said to have a couple of disadvantages as follows:

1.  Drain installation requires a large workforce; four to five workers are required to construct each drain
2.  Sand drains also need suitably graded sands. In some cases, these sands may have to be imported from other places. This may cause the cost of the drains to be very high.

As an alternative to sand drains, prefabricated strip drains are increasingly used because they are easier and cheaper to install. The technique for installation of these prefabricated drains may vary depending on the equipment used as well as specialist contractors available for their installation. But the general principle of drain installation can be described as follows.

Figure 4.22 shows a schematic of prefabricated vertical drain installation into the subsoil by using a mandrel (square cross-section steel tube) driven by a specially adapted machine such as a dragline or a pile driver. The prefabricated vertical drain, which comes in coil form, is placed on the machine. The base of the drain is fitted with a special "shoe" made either of steel or plastic, thereby covering the base of the mandrel during the driving operation. The

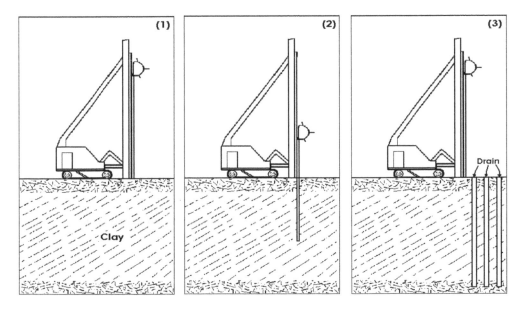

*Figure 4.22* Schematic diagram of prefabricated vertical drains installation

shoe then functions to anchor the drain to the soil upon reaching the desired depth of installation as the mandrel is being pulled out.

These drains can be installed to depths of 40 m or more. An installation machine together with a crew of two workers may install up to 4000 m length of drain per day.

In general, there are two methods of driving the mandrel: the jetting and dry vibrating technique. It is debatable which method is more appropriate. In any case, with the dry vibrating method, the soil surrounding the drain will be remolded. This will give an adverse effect to the soil shear strength and its consolidation characteristics. Because of this, the jetting method seems to be more suitable. But the quantity of water needed for this case needs to be limited in order to minimise possible environmental contamination of the site.

To achieve the desired degree of average consolidation within the time specified, preloading by surcharge technique with or without the loading berms may also be used. This can reduce the number of drains required. The surcharge will be removed after the required consolidation settlement under working load has been achieved.

### 4.4.5   Horizontal drains

#### 4.4.5.1   Design of horizontal drains

Horizontal drains are more suitable for shallow stabilisation, such as for natural slopes and along the highways. This method is not suitable for deep stabilisation because of its depth limitation, which is only 5–7 m. This is because there is no suitable machine that can dig very deep in addition to the consideration of side instability. The drains are usually in the form of trenches backfilled with granular material (sand and gravel) or a combination of perforated pipes with the sand and gravel.

Calculating spacing and depth of drains to achieve the desired degree of consolidation is usually done using analysis based on two dimensional consolidation theories. The quantity of water flowing into the drain can be estimated using the flow net method.

### 4.4.6   Horizontal drain installation method

Construction of horizontal drain is made by using special equipment, by digging up trenches and then backfilling them with granular materials.

### 4.4.7   Electro-osmosis

Casagrande first introduced the method of treating soft soil with electro-osmosis in 1947. Electro-osmosis involves water transport through a continuous soil particle network, where the movement is primarily generated in the diffuse double layer or moisture film where cations dominate. When the direct electrical gradient is applied to a clay-water system, the surface or particle is fixed, but the mobile diffused layer moves and the solution with it is carried. The main mechanism in electro-osmosis is the migration of ions, meaning the cations migrate to the cathode and the anions move toward the anode (Figure 4.23). However, this method is not used in large-scale projects because of its high cost.

The electro-osmotic flow that results from the fluid surrounding the soil particles is induced by ionic fluxes. In addition, the water molecules in flow in bulk phase can be carried out along with flow from the fluid surrounding the soil particles in the same flow

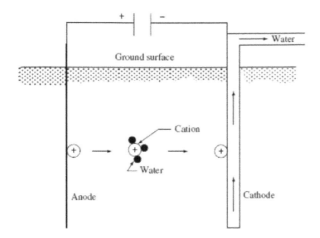

*Figure 4.23* Principles of electro-osmosis

Source: Adapted from Das (2008)

direction. Interaction between flow in the fluid surrounding the soil particles as a first region and flow in the bulk phase as a second region enables the movement of water in the bulk phase, meaning a drag action is the main cause of the electro-osmotic flow. Therefore, the total observed electro-osmotic flow is attributed to the movement of these two water layers (Asadi et al., 2013).

The Helmholtz-Smoluchowski theory is one of the widely used models to describe electro-osmotic processes. The Helmholtz-Smoluchowski theory assumes the pore radii are relatively large in comparison to the thickness of the diffuse double layer and the mobile ions are concentrated near the soil-water interface (Figure 4.24). Based on the Hemholtz-Smoluchowski theory, the zeta potential ($\zeta$) and the charge distribution in the fluid adjacent to the soil surface play important roles in determining the electro-osmotic flow.

The rate of electro-osmotic flow is controlled by the coefficient of electro-osmotic permeability of the soil, $k_e$, which is a measure of the fluid flux per unit area of the soil per unit electric gradient. The value of $k_e$ is a function of the zeta potential ($\zeta$), the viscosity of the pore fluid, the soil porosity and the soil electrical permittivity. The coefficient of electro-osmotic permeability is given by Equation (4.24):

$$q = \frac{\varepsilon \zeta}{V_t} n \frac{E}{L} A \qquad (4.24)$$

where

$\zeta$ = zeta potential
$V_t$ = viscosity of the pore fluid
$n$ = soil porosity
$\varepsilon$ = soil electrical permittivity
$A$ = gross cross-sectional area perpendicular to water flow
$L$ = length
$q$ = flow rate.

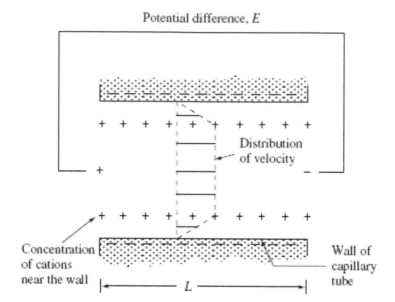

*Figure 4.24* Helmholtz-Smoluchowski theory for electro-osmosis

Source: Adapted from Das (2008)

The hydraulic conductivity, $k_h$, is significantly affected by the pore size and distribution in the medium, but $k_e$ based on the Helmholtz-Smoluchowski theory is dependent mainly on $\zeta$ and $n$.

The concentration of electrolytes, type of electrolytes, valance of ions and pH are important factors that can affect $\zeta$ values.

Das (2008) reported that Schmid in 1951 proposed a theory in contrast to the Helmholtz-Smoluchowski theory. It was assumed that the capillary tubes formed by the pores between clay particles are small in diameter, with the result that excess cations would be uniformly distributed across the pore cross-sectional area (Figure 4.25). Based on this theory, we have Equation (4.25):

$$q = n \frac{r^2 A_0 F}{8 V_t} \frac{E}{L} A \tag{4.25}$$

where

  $r$ = pore radius
  $A_o$ = volume charge density
  $F$ = Faraday constant
  $n$ = porosity
  $A$ = gross cross-sectional area perpendicular to water flow
  $L$ = length
  $V_t$ = viscosity
  $q$ = flow rate.

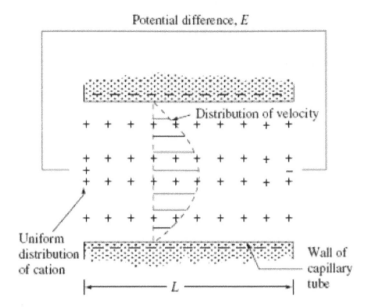

*Figure 4.25* Schmid theory for electro-osmosis

Source: Adapted from Das (2008)

However, the most widely used electro-osmotic flow equation for the soil system is proposed by Casagrande (1949) (Equation 4.26):

$$q = k_e i_e A \qquad (4.26)$$

where

$A$ = gross cross-sectional area perpendicular to water flow
$i_e$ = applied electrical gradient
$k_e$ = coefficient of electro-osmotic permeability
$q$ = flow rate.

Rate of water flow in the electro-osmosis treatment method can also be estimated using Bell's (1975) equation as follows:

$$q = k_e \rho I \qquad (4.27)$$

where $q$ is the rate of water flow, $k_e$ is the efficiency of the electro-osmosis (typical value 0.005 m²/Vs at moisture content 60%), $\rho$ is soil resistance and $I$ is the current used.

Equation (4.23) shows that the rate of water discharge is directly proportional to the current, $I$, and soil resistance, $\rho$. In practice, the value of $k_e$ reduces with a reduction in water content. Therefore, the effectiveness of the method is also reduced. To improve this situation, electro-osmosis can be combined with chemical treatment to accelerate absorption of ions in soils of low permeability.

The method is effective for the following soil conditions:

- Silt or saturated clayey silt
- Normally consolidated soil
- Pore water with low electrolyte intensity
- A potential gradient in the same direction as the hydraulic gradient.

### 4.4.8   Electro-osmosis treatment installation method

The equipment required for the electro-osmosis treatment and the installation procedures are actually very simple. They involve inserting steel or aluminum rods of diameter 1–10 cm or perforated pipes into the soil to act as the anode. Wells are usually used as the cathode. Direct current with potential gradient between 20 V/m and 50 V/m, depending on types of soil, is then applied.

The depth of soil treated with this method is between 10 m and 20 m.

## Case Studies

### Case 1: Evaluation of treatment methods used for construction on expansive soils in Egypt

The case study is for a site located in the Katamia region (east of Cairo) where swelling soil was found. In this site, five boreholes were dug around the main building, in which more cracks appeared in its concrete structures (beams and columns), to investigate the underlying soil properties. It was found that the swelling soil formation was when the soil was wetted. Thus, dry samples were taken from additional test pits which were dug in un-wetted areas near the defective main building to perform the laboratory swelling pressure test. The swelling soil pressure was very high ranging from 6 kg/cm$^2$ to 12 kg/cm$^2$ for dry soil and the soil bulk density ranged from 1.96 t/m$^3$ to 2.00 t/m$^3$. Free swelling index readings were in the range of 110%–170%. Figure 4.26 illustrates the value of the laboratory swelling pressure achieved on the dry soil samples (two samples were shown) which were collected from test pit excavation near the defective building.

The defective structure was a governmental building having nearly 1000 workers and consisted of three adjacent building parts with a structural joint between each part. The shallow foundation type was an inverted T-beam (shallow strip footings). In this site, the soil replacement was sand cushion material which was backfill just under each strip footing without any extension width outside the footing area from each side, as shown in Figure 4.27.

Also, there was no backfill soil under and around the slab. And due to the lack of sand cushion of the soil replacement in the area between the strip footings, the swelling soil in this area was still active and applied upward pressure on the foundation and the slab above, especially when there was water leakage from the adjacent water tank, which was 2 m distant from the main building. And thus, with the change in the underlying soil

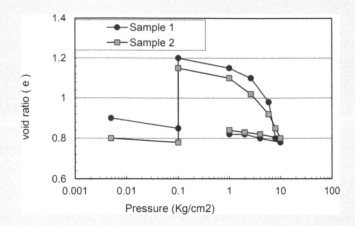

Figure 4.26 Swelling pressure test

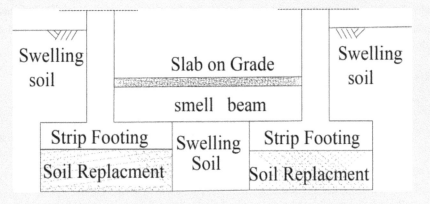

Figure 4.27 Soil replacements under strip footing

moisture content, the expanded soil in volume moved the footings up and caused the cracks and the horizontal movement shown between the parts of the main building. The horizontal movement was observed each floor and increased at the top of the building. Also, cracks were observed at the different structural elements. By investigating the expansive soil under and around the main governmental building, it was found in a wet condition, and this was due to water leakage from the nearest groundwater tanks and the irrigation around the building directly. Also it was found that the location of the main building was in a depressed location of the whole site, which enabled the rainwater and leakage from the groundwater tank to be collected under its footings, which was not surrounded with enough replacement soil.

Authors: Farid and Amin; http://www.jeaconf.org/UploadedFiles/Document/d4b825c2-3d29-460b-ba39-0e89b03e5a4e.pdf

## Case 2

(Ground improvement using PVD with preloading for coal and iron ore stackyard) (Source: Bhosle and Vaishampayan, Proc., IGC 2009, India)

The development of the new port at Gangavaram located about 15 km south of Visakhapatnam Port, Andhra Pradesh, commenced in December 2005. The preliminary soil investigation revealed the presence of soft clay up to a depth of 10–18 m with very low safe bearing capacity and high consolidation parameters. It was therefore decided to enhance the soil properties using a ground treatment method: "Use of Band Drains/PVD with Pre-loading." In general, the area was fairly level and a thin layer of dredged sand of thickness 0.2–0.3 m was present at most of the location. Immediately below the dredged sand was the marine clay with shells with thickness of 1–3 m. A layer of soft marine clay of thickness 7–15 m was observed following this layer of marine clay with shells. The standard penetration test was conducted in this stratum at various depths, indicating the penetration of 45–60 cm in one blow thus the N resistance was 0–1. At a depth of 12–18 m below the existing ground level, the penetration resistance N was observed to be increasing with depth. The safe bearing capacity of the existing soil was worked out as 3 ton/m², which was very low to take the loads. The consolidation settlements were worked out at 1000–1600 mm.

### Salient Features of Scheme

Machinery used: Hydraulic stitchers

Depth of PVD: 10–18 m below original ground level (OGL)

Spacing of PVD: 1 m center to center (c/c) in a triangular grid below the stacker reclaimers and 1.5 m c/c in a triangular grid in other areas

Consolidation period: For 1 m spacing, 65 days; for 1.5 m spacing, 174 days

Thickness of sand mat: 300 mm

Horizontal drainage system: Combined system of geotextile pipe formed by boulders/gravel encased in geotextile and band drains laid horizontally connecting these pipes.

Table 4.1 shows the details of the instruments installed in the project area.

Figures 4.28 and 4.29 indicate the excess pore pressure variations with time. For the piezometers installed below the stacker reclamation, the degree of consolidation was

*Table 4.1* Details of instrument

| Section | Casagrande piezometer | Vibrating wire piezometer | Plate settlement markers | Magnetic settlement recorder |
|---------|-----------------------|---------------------------|--------------------------|------------------------------|
| Section 1 | CP1 | VP1, VP2, VP3 | PS1. PS2 | MS1, MS2 |
| Section 2 | CP4 | VP9, VP 10, VP 11 | PS8. PS9. PS10 | MS6 |
| Section 3 | CP3 | VP6, VP7, VPS | PS5, PS6, PS 7 | MSS |
| Section 4 | CP2 | VP4, VP5 | PS3, PS4 | MS3, MS4 |
| Section 5 | CP5 | VP 12, VP 13, VP 14 | PS11, PS 12, PS 13 | MS7 |

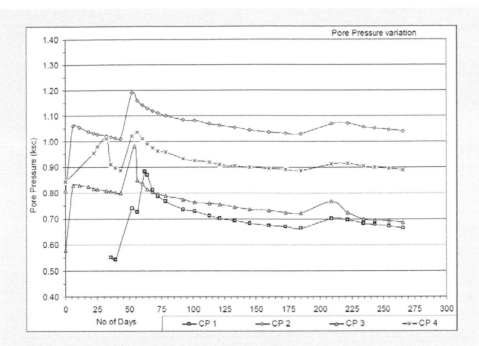

Figure 4.28 Variation of excess pore pressures with time

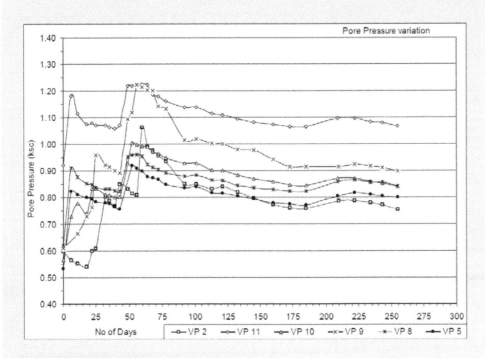

Figure 4.29 Variation of excess pore pressures with time

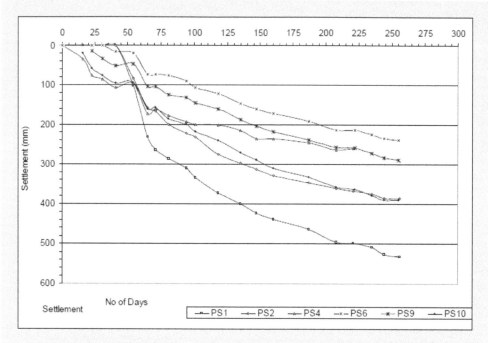

*Figure 4.30* Settlements with time

observed as 55%–65%. For the piezometers installed in other area (PVD spaced at 1.5 m c/c), the degree of consolidation was observed as 30%–40%.

Figure 4.30 indicates the settlement data with time. The observed settlements were in the range of 300–500 mm.

## Case 3

Ground improvement using preloading with prefabricated vertical drains (Source: Gadhiya Shadab and Vyas Saurabh, 2013, Indian Geotechnical Conference)

Essar Steel Orissa Ltd. has proposed to set up an 8 MTPA integrated iron ore pellet/steel plant at Paradeep, which comprises various facilities and stockpile areas where the movement of stacker-cum-reclaimer was planned. Heavy loading in terms of metric tons was expected and the subsoil beneath consisted of soft soil in nature.

The area under study is generally flat. The study area is primarily deltaic alluvial sediments drained by the Mahanadi River near its confluence with the Bay of Bengal. The topography of the area is such that it consists of low-lying areas and a nearby creek.

A total of 13 boreholes were drilled to gather subsoil information for the site. The boreholes were distributed over the area; 10 boreholes were sunk to 50 m depth and three boreholes were sunk to 30 m depth. Seven vane shear tests, four static cone penetration tests (SCPT) and three field CBR tests were also carried out at specified locations. Observed SPT values of boreholes in the stockyard area are shown in Figure 4.31.

Looking to the soil stratification and properties, the top few meters of soil were too weak to resist the load of stacking, and the actual height of stacking was undetermined at the time of proposal. So it was decided to stack 1.5 m height of preload with locally available soil. Considering 90% of degree of consolidation to be achieved, the design of ground improvement with PVD was done.

The time required for 90% consolidation was observed to be 3.87 years. This period is very much higher than practical considerations. Hence, the installation of PVDs was considered to be a viable option to accelerate the consolidation process. Calculation showed

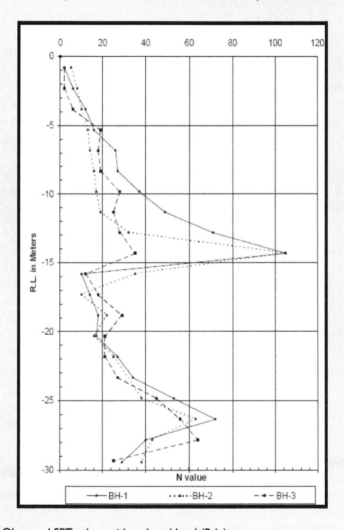

*Figure 4.31* Observed SPT values with reduced level (R.L.)

*Table 4.2* Properties considered for design of embankment

| Layer | Unit Weight (kN/m³) | | | C or Cu (kPa) | | | φ (degree) | | |
|---|---|---|---|---|---|---|---|---|---|
| | Product Pellet | LC | DC | Product Pellet | LC | DC | Product Pellet | LC | DC |
| Preload | 18 | 17.85 | 18 | 0 | 19 | 23 | 30 | 0 | 0 |
| I | 22 | 17.85 | 18 | 0 | 34.91 | 39.1 | 38 | 0 | 0 |
| II | 22 | 17.85 | 18 | 0 | 50.81 | 55.08 | 38 | 0 | 0 |
| III | 22 | 17.85 | 18 | 0 | 60 | 68 | 38 | 0 | 0 |

LC – Soft brownish gray to gray silty clay 1.5 m thick.
DC – Loose to medium dense/dense grey silty clay 6 m thick.

that the time for 90% could be reduced to 1 month by installing PVDs. Triangular grid spacing of 1.2 m was suggested.

Product pellets of 100 × 40 × 8 m is considered, with 20 m space in between. Looking to the movement of stackers and reclaimers in the stockyard area, triangular stacking in three stages was assumed. The height of each stage of stack is 3 m, 3 m and 2 m, respectively. Table 4.2 shows the properties considered for design of embankment at different stages.

The embankment was modeled in SETTLE 3D software for settlement computation. The complete cycle of preloading, removing and actual stages of loading was modeled assuming the pressure distribution as 2:1. The estimated settlement was 1.02 m.

## Case 4

Case study on hydraulic reclaimed sludge consolidation using electrokinetic geosynthetics (EKG) (Zhuang et al., 2014).

Currently the most popular technique for consolidating hydraulic-filled areas is vacuum preloading. However, for hydraulic reclaimed sludge, vacuum preloading is too slow. Also for area of deep hydraulic reclaimed sludge (usually deeper than 4–5 m), the effect of vacuum preloading is limited. In this case, electro-osmotic consolidation can be an alternative option.

There was a 19 m × 15 m hydraulic reclaimed area to be treated by electro-osmotic consolidation. The area was filled with 5.8 m thick dredged pool sludge. The properties of the sludge before electro-osmotic consolidation are shown in Table 4.3.

EKG electrodes were square, arrayed at a space of 1 m. The treatment included two stages separated by 16 days of intermittence. In stage 1, electro-osmotic consolidation began under a constant-current mode of 290 A and lasted for 233.57 hrs (~ 10 days). Then it was switched to a constant-voltage mode of 50 V and lasted for 28.55 hrs (~ 1 day). In stage 2, after the intermittence, electro-osmotic consolidation continued under a constant-voltage mode of 80 V and lasted for 215.02 hours (~ 9 days). The treatment was controlled by computer software. Electrode polarity was reversed according to the trend of electric current variation. Generally,

Table 4.3 Properties of sludge before electro-osmotic consolidation

| Specific gravity | Water content (%) | Dry density (g/cm³) | Permeability (cm/s) | Liquid limit (%) | Plastic limit (%) |
|---|---|---|---|---|---|
| 2.61 | 62 | 1.03 | 3.0 × 10⁻⁷ | 50 | 22 |

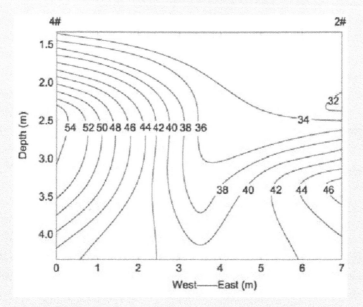

Figure 4.32 Distribution of water content after electro-osmotic dewatering along middle cross-section from the west to the east (within borehole range)

stage 1 had a longer time of current in reverse direction, while stage 2 had a longer time of current in forward direction. The purpose of this scheme for electro-osmosis was to maintain mobile cations in the soil as much as possible.

After the electro-osmotic dewatering treatment, soil was sampled from the field for measurement of properties in the lab. Test results showed that the water content of the soil decreased from an average of 62% to 36% and the minimal water content after electro-osmotic dewatering according to the borehole was 24%. Distribution of water content after electro-osmotic dewatering along the middle cross-section from west to east and from north to south is shown in Figures 4.32 and 4.33, respectively. Water content was relatively high to the east and low to the south. This is due to the boundary conditions. There was a pond at the east, while there was a road at the south.

Unconsolidated-undrained shear strength of the soil increased from 0 to 25 kPa; the soft ground was improved from a fluid-like status to a bearing capacity of 70 kPa. The average energy consumption for this treatment was 5.6 kWh/m.

As a comparison, the preloading method was analysed to produce the same consolidation effect as EKG. The soil had a coefficient of consolidation CV = 0.0029 cm²/s and a

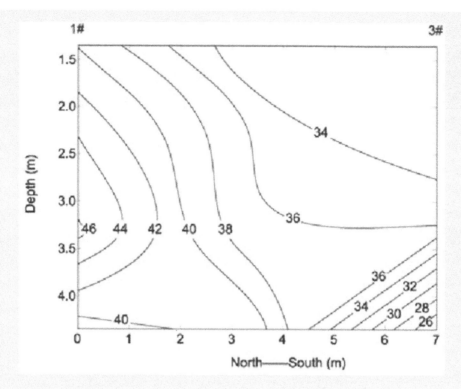

*Figure 4.33* Distribution of water content after electro-osmotic dewatering along middle cross-section from the north to the south (within borehole range)

compression index $C_c = 0.3611$. In order to achieve the same effect of consolidation (to reduce water content from 62% to 36%), there should be 132 kPa of preloading (around 6–7 m high soil surcharge); and it would take 1139 days – more than three years – to achieve 90% of consolidation. However, electro-osmotic consolidation took only 36 days, including 16 days of intermittence.

For vacuum consolidation, the theoretic limit of bearing capacity is 1 atm (~100 kPa) and the practical limit is around 80 kPa considering the vacuum loss. Therefore, practical vacuum consolidation can achieve a bearing capacity of 50–60 kPa in 3–6 months for hydraulic reclaimed sludge. The comparisons above show that electro-osmotic consolidation is much quicker and can achieve better consolidation effect.

This case study shows that if the electro-osmosis scheme is properly designed, the energy consumption is fairly acceptable. It is not higher than the energy consumption for vacuum consolidation. The high cost is due to the price of the EKG product. The EKG product adopted in the field application is 10 times more expensive than traditional PVD. EKG can compete with PVD in consolidation projects only if the price of EKG decreases or the effect of consolidation further improves so that the secondary treatment of the ground is not necessary after electro-osmotic consolidation.

# References

Ali, F. & Abdul Wahab, M. T. (2003) The geotechnical assessment on construction of coastal protection structure on soft sea bed. In: Huat et al. (eds) *Proceedings of 2nd International Conference on Advances in Soft Soil Engineering and Technology*. Universiti Putra Malaysia Press, Putrajaya, Malaysia. pp. 755–763.

Ali, F. H. & Huat, B.B.K. (1992a) Improvement of soft soil using preloading and vertical drainage. *Proceedings of Regional Seminar on Land Reclamation for Urban Development, Kuala Lumpur, August 10–12*. University of Malay, Kuala Lumpur.

Ali, F. & Huat, B.B.K. (1992b) Performance of composite and monolithic prefabricated vertical drains. *Pertanika*, 15(3), 255–264.

Asadi, A., Huat, B.B.K., Nahazanan, H. & Keykhah, H. A. (2013) Theory of electroosmosis in soil. *International Journal of Electrochemical Science*, 8, 1016–1025.

Barron, R. A. (1948) Consolidation of fine-grained soils by drain wells. *Transactions, ASCE*, 113, 718–754.

Bell, F.G. (1975) *Methods of Treatment of Unstable Ground*. Butterworth & Co. Ltd., London.

Bishop, A. W. (1959) The principle of effective stress. *Teknisk Ukeblad*, 106(39), 859–863.

Bhosle, P. & Vaishampayan, V. V. (2009) Case study for ground improvement using PVD with preloading for coal & iron ore stackyard. *Proceedings of Indian Geotechnical Conference, 17–19 December, Guntur, India*. Indian Geotechnical Society, Guntur, India.

Carlsten, P. (1988) Peat geotechnical properties and up to date methods of design and construction. *Proceedings of 2nd Baltic CSMFE*. Swedish Geotechnical Institute, Tallinn.

Carillo, J. J. (1942) Simple two and three dimensional cases in the theory of consolidation, *Journal of Mathematics and Physics Banner*, 21(1), 1–5.

Casagrande, I. L. (1949) Electro-osmosis in soils, *Géotechnique*, 1(3), 159–177. doi:10.1680/geot.1949.1.3.159

Das, B. M. (2008) *Advanced Soil Mechanics*. Taylor & Francis, New York. p. 594.

Edil, T. B. (1994) Immediate issues in engineering practise. In Den Haan et al. (eds) *Proceedings of Conference on Advances in Understanding and Modelling the Mechanical Behaviour of Peat*. Balkema, Rotterdam, The Netherlands. pp. 403–444.

Gadhiya Shadab, A. & Vyas Saurabh, D. (2013) Ground improvement using pre-loading with prefabricated vertical drains: A case study. *Proceedings of Indian Geotechnical Conference, 22–24 December, Roorkee, India*. Roorkee.

Han, K. K. & Eng, S. H. (2003) Performance of ground improvement at Vung Tau, Vietnam. In Huat et al. (eds) *Proceedings of 2nd International Conference on Advances in Soft Soil Engineering and Technology, July 2003*. Universiti Putra Malaysia Press, Putrajaya. pp. 601–608.

Indraratna, B., Rujikiatkamjorn, C., Balasubramaniam, A. S. & Wijeyakulasuriya, V. (2005a) Predictions and observations of soft clay foundations stabilized with geosynthetic drains and vacuum surcharge. In: Indraratna, B. & Chu, J. (eds) *Ground Improvement – Case Histories Book*, Volume 3. Elsevier, London. pp. 199–230.

Indraratna, B., Rujikiatkamjorn, C. & Sathananthan, I. (2005b) Radial consolidation of clay using compressibility indices and varying horizontal permeability. *Canadian Geotechnical Journal*, 42, 1330–1341.

Indraratna, B., Rujikiatkamjorn, C., Ameratunga, J. & Boyle, P. (2011) Performance and prediction of vacuum combined surcharge consolidation at Port of Brisbane, *Journal of Geotechnical & Geoenvironmental Engineering, ASCE*, 137(11), 1009–1018.

Jarrett, P. M. (1997) Recent developments in design and construction on peat and organic soils. In: Huat & Bahia (eds) *Proceedings of Conference on Recent Advances in Soft Soil Engineering*. Samasa Press Sdn. Bhd., Kuching, Sarawak. pp. 1–16.

Jennings & Burland (1962). Limitations to the use of effective stresses in partly saturated soil, *Géotechnique*, 12(2), 125–144.

Kjellman, W. (1952) Accelerating consolidation of fine grained soils by means of card wicks. *2nd International Conference on SM&FE*, Rotterdam, The Netherlands.

Kurihara, N., Isoda, T., Ohta, H. & Sekiguchi, H. (1994) Settlement performance of the central Hokkaido expressway built on peat. In: Den Haan et al. (eds) *Proceedings of conference on Advances in Understanding and Modelling the Mechanical Behaviour of Peat*. Balkema, Rotterdam, Brookfield. pp.361–367.

Laughton, A. S. (1955) *The Compaction of Ocean Sediments*. Ph.D. Thesis, University of Cambridge, England.

Mitachi, T., Yamazoe, N. & Fukuda, F. (2003) FE analysis of deep peaty soft ground during filling followed by vacuum preloading. In: Huat et al. (eds) *Proceedings of 2nd International Conference on Advances in Soft Soil Engineering and Technology*. Universiti Putra Malaysia Press, Putrajaya, Malaysia. pp.267–276.

Osterberg, J. O. (1957) Influence values for vertical stresses in a semi-infinite mass due to an embankment loading. In: *Proceedings of Fourth International Conference on Soil Mechanics and Foundation Engineering*. Butterworth, London.

Rowe, P. W. (1972) The relevance of Soil Fabric to Site Investigation Practise. *12th Rankine Lecture, Geotechnique*, 22(2), 195–300.

Rujikiatkamjorn, C. & Indraratna, B. (2013) Current state of the art in vacuum preloading for stabilising soft soil. *Geotechnical Engineering Journal of the SEAGS & AGSSEA*, 44(4), 77–87.

Skempton, A. W. (1960) Effective stress in soils, concrete and rocks. *Pore Pressure Conference*. Butterworth, London.

Terzaghi, K. (1925) Principles of soil mechanics. *Engineering News-Record*, 95(19–27), pp. 742–746, 796–800, 832–836, 874–878, 912–915, 987–990, 1026–1029, 1064–1068.

Van den Elzen, L.W.A. & Atkinson, M.S. (1980) *Accelerated consolidation of compressible low permeability subsoils by means of Colbond Drains*, Arnheim, Colbond b. v.

Zhuang, Y. F., Huang, Y., Liu, F., Zou, W. & Li, Z. (2014) Case study on hydraulic reclaimed sludge consolidation using electrokinetic geosynthetics. *10th International Conference on Geosynthetics, 21–25 September 2014*. DGGT, Berlin, Germany. (CD-ROM).

# Chapter 5

# Fibers and geosynthetics

## 5.1 Introduction

In construction practice, fiber-reinforced soil has become a viable, cost-effective and environmentally friendly ground improvement technique to improve stability and control deformation in various applications including retaining structures, embankments, foundations, slopes and pavements. To be effective, the reinforcements must intersect potential failure surfaces in the soil mass. Strains in the soil mass generate strains in the reinforcements, which in turn generate tensile loads in the reinforcements. The technique of reinforced soil with flexible, discrete fibers is not a new concept. Many ancient structures incorporated layers of natural tensile elements to reinforce the soil for construction of stable structures. The ancient Romans, for example, used natural materials in the form of woven reed mats to assist road construction over soft soils. In the Far East, there were earth structures reinforced with bamboo or reeds.

In the last three decades or so, the use of synthetic fabrics (called geosynthetics) both woven and non-woven, have been increasingly used in civil engineering constructions along with several other applications of specific geosynthetics to achieve different functions, such as separation, filtration, drainage, fluid barrier and protection. Geosynthetic reinforcements include woven geotextiles, geogrids, some geocomposites and so forth. In noncritical structures, natural products (also known as geonaturals) are also used as soil reinforcements (bamboo, geocoir, geojute). It is not really known when the first synthetic fabric was actually used. Nevertheless, it can be said that the usage of modern synthetic fabric began in Europe in the mid-1960s. These earlier fabrics were mainly used in constructions such as material separators for roads, for coastal erosion control, for slope and pavement reinforcement and for underground soil drainage systems.

In Malaysia, among the first usage of fibers and geosynthetics was for road embankment construction, both for separation and reinforcement, as described by Huat and Aziz (1988). Nowadays, impermeable fabrics (or synthetic membranes) are also available for use in reservoirs, linings for tanks and other water-filled structures. This chapter describes the research and field applications of fiber-reinforced soils focusing on presenting research to practice in the region.

As mentioned, soil reinforcement is a procedure where natural or synthetised additives are used to improve the properties of soils. Several reinforcement methods are available for stabilising problematic soils. In this chapter, mechanical soil reinforcement method using fibrous material is described. Figure 5.1 presents a state of review of mechanical procedures of soil reinforcement using fibrous materials.

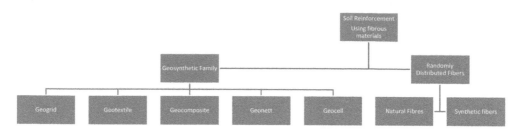

*Figure 5.1* Soil reinforcement using fibrous material

## 5.2  Fiber-reinforced soil

In most applications, discrete fibers are simply added and mixed randomly with soil or other similar materials (coal ash, mine tailings, etc.). Fibers can either be mixed through the soil matrix material manually or by mechanical means. The mechanical procedure can be divided into three categories: cultivator mixing, concrete mixer and tumbler mixer.

Many published experimental studies implicitly assume that the fibers are randomly oriented throughout the soil mass (Anggraini et al., 2016a). Such a distribution of orientation would preserve the soil strength isotropy and eventually avoid or delay formation of localised deformation planes. However, it has been found that the most common procedure for preparing reinforced specimens, tamping, leads to preferred sub-horizontal orientation of fibers. Soil reinforcement with fiber (natural and synthetic) has been and is still a popular and frequently used approach which is globally applied, as discussed by various researchers (Freitag, 1986; Maher and Ho, 1994; Prabakar and Sridhar, 2002; Kaniraj and Gayathri, 2003; Park and Tan, 2005; Anggraini et al., 2017; Mirzababaei et al., 2018). The application of synthetic fibers such as polyvinyl alcohol (PVOH), polypropylene (PP) and glass has also been extensively discussed (Park, 2011; Musenda, 1999; Consoli et al., 2004). Recent findings have also indicated the possibility of the application of nylon, polyethylene and steel fibers, which have high tensile strength and durability.

The utilisation of natural fibers as reinforcement materials has recently gained tremendous attention around the world due to its availability, sustainability, and environmentally friendly nature (Prabakar and Sridhar, 2002; Abtahi et al., 2010; Marandi et al., 2008; Lin et al., 2010; Lekha et al., 2015; Anggraini et al., 2016a; Sivakumar et al., 2008). Natural fiber materials include coconut (coir), bamboo, jute, sisal and palm. The effectiveness of natural fibers in soil reinforcement depends on the fiber strength as well as its interaction with the soil. Moreover, the behavior of the reinforced soil structure is also associated with both the soil and fiber properties. Apart from the individual properties of both the soil and reinforcement material, the interaction between reinforcement material and soil also plays an important role in deciding the behavior of reinforced soil structure. Consequently, when a tensile force needs to be mobilised in fibers, adhesion restrains the fibers from pulling out and thus allows tensile resistance to develop. A study by Anggraini et al. (2016b) indicated that chemical treatment of coir fiber by $CaCl_2$ and NaOH improves the tensile strength of the fiber significantly by 211% as compared to untreated coir fiber, and consequently increases the strength performance of the fiber-reinforced marine clay. From their study, it can be acknowledged that the interface roughness of fiber and soil play an important role in the reinforced soil system.

SEM analysis was conducted by Anggraini et al. (2016b) to show the interaction of fiber-reinforced soil and lime used as an additional material. It is observed that the white dots

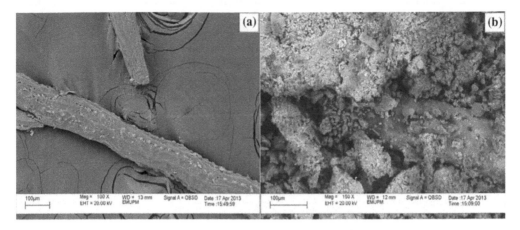

*Figure 5.2* SEM of surface aspect of single fibers (a), fibers in soil (b)

Source: Anggraini et al. (2015)

found on the surface of the single fiber which can be seen in Figure 5.2 were identified as silica-rich material, which in its fine state does react with quicklime (Bell, 1996). As can be seen in Figure 5.2, lime-treated marine clay has fabric with cementitious gel between aggregate soil particles. In addition, some pores between particles are filled with cementitious gel formed, which also results in fabric with small pores in lime-treated soil. Hence, soil treated with lime contributes to a denser soil fabric. These reaction products bind the soil particles around them together and strengthen the soils.

For coir fiber and lime-treated soft soil, the pure fiber soil has fabric with packets of soil particles. It may be because of the presence of water; the clay particles adhered to each other or formed packets of soil particles. This was attributed to the better interfacial adhesion of coir fiber-reinforced cemented soil due to the increase in fiber roughness and contact area of fibers. Some parts of the fiber were pulled out from the soil when shearing occurred and the fiber itself was not sheared off. From the abrasion trace in the fiber surface, it is indicated that the fiber strengthened the soil by the friction between fiber and soil. Therefore, the combined fiber and lime inclusions increase the efficiency of transfer of the load from the matrix to the fibers.

This natural fabric can also be used as aid for construction on soft soils. Ramaswamy et al. (1982) reported studies carried out at the National University of Singapore on the usage of jute fabric for reinforcing a road embankment on soft soil. However, the main criticism for the use of natural material such as jute is that it lacks long-term durability. It cannot be denied that jute will rot easily when left in the soil environment for too long. Having said that, it must be appreciated that even natural materials such as jute will initially function as a strong reinforcing material. With time, the soil will consolidate through the process of dissipation of excess pore water pressure generated by the weight of the overburden material, pavement and traffic. This will cause the soil strength to increase through the process of consolidation. If this gain in strength of the founding soil is proportional to the loss strength of the jute fabric, the soil will therefore become less dependent of the fabric in the long term. The long-term lack of durability aspects of jute will therefore become less important.

Durability is a main problem involved with using fibers in soil reinforcement. Attempts are being made to increase the long-term durability of fibers in a cost-effective way, such as coating fibers with phenol and bitumen. However, natural fibers can be used normally in less critical applications such as pavement bases/sub-bases or short-term applications such as erosion control.

Randomly distributed fibers (natural and/or synthetic fibers) in geotechnical engineering are feasible in six fields including pavement layers (road construction), retaining walls, railway embankments, protection of slopes, and earthquake and soil-foundation engineering. In all applications, the discrete fibers are simply added and mixed randomly with soil or other similar materials with and/or without additives. However, reinforcing soil with short fibers is a ground improvement technique that has not yet been fully utilised worldwide, especially in Southeast Asia.

The laboratory investigation and numerical analysis–based investigation explored the effects of treated short fibers and lime on the mechanical performance of the treated soil as a pile supported load platform was investigated by Anggraini et al. (2015). A diagram of the model test is given in Figure 5.3. The model is 800 mm wide, 400 mm long and 300 mm

*Table 5.1* Types of materials for making of fabric

|  | Fiber materials |
| --- | --- |
| Natural | Coconut (coir), bamboo, jute, sisal, palm |
| Synthetic (fabrics) | Polyamides (nylon) |
|  | Polyester (PET) |
|  | Polyolefin (polypropylene, polyethylene) |
|  | Polyvinyl (PVA, acrylic) |
|  | Polystyrene |

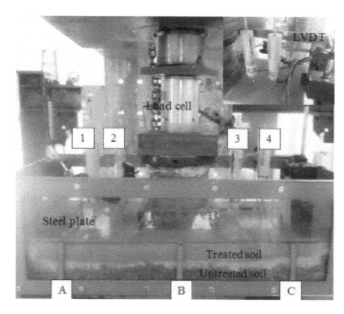

*Figure 5.3* Test setup and instrumentation detail of the physical model of an earth platform in the laboratory platform

Source: Anggraini et al. (2015)

high. The unit weight of this material was determined as 16 kN/m³. The soft soil is simulated by 100 mm thick untreated soft soil. These elements aim to simulate a real subsoil layer and settlements at the platform base are obtained. The effect of the soft subsoil is rarely taken into account, whereas the mechanisms developing above the piles and in the soft subsoil are connected (Jenck et al., 2007). By placing sand in alternate color layers, the platform settlement was observed under a maximum applied load. Four load transfer transducers are placed on the platform, which permits quantification of the differential settlement occurring between the rigid piles and the soft soil.

Two units of circular steel bar are used to represent the piles, which are fixed to the rigid apparatus frame to avoid any vertical and lateral displacements of the piles. The mechanisms at the central zone of the earth platform between two piles assume that no boundary effect is observed. The platform is set up as a 50 mm thick layer. A 112.5 kPa surcharge application constituted by the actuator is then placed at the surface. The settlements were recorded at 6.25 kPa surcharge increments. The developed model presents modularity in terms of geometrical parameters.

However, this physical model presents limitations, which are as follows:

1.   The model proposed is a simplification of the reality as it considers a two-dimensional case, whereas this type of system is typically three-dimensional.
2.   The similarity rules are not strictly respected. However, this physical model does not aim at simulating the behavior of a real system but it is used to understand the efficacy of an earth platform in order to determine the effectiveness of reinforcement in reducing settlement and enhancing bending performance.
3.   The aim of this study is to observe the behavior of the earth platform; the behavior of the soft soil below the earth platform was not taken into account (i.e., bending, end bearing and friction of pile). The length of the pile was not considered; it is used as a support in order to study the flexural behavior of the earth platform.
4.   In this model, geometric scaling was adopted. Settlement beneath the earth platform induced by the 112.5 kPa surcharge application was measured. However, the total force applied will have the same value for the prototype and the scaled-down model when the geometry is reduced.
5.   Platform material properties cannot be varied in the small-scale model. Hence, cohesion, friction angle, Young's modulus and Poisson's ratio were not scaled down.
6.   In this model, the earth platform height is generally limited and is not enough to develop an arching mechanism.

The numerical analyses showed good agreement of the importance of the mechanical properties of fiber-reinforced soil for the efficacy and effectiveness of the reduction of the settlement of the earth platform, as well as to enhance the bending performance of the earth platform as shown in Figure 5.4. The differential settlement at the elevation of the pile heads is significantly reduced by an increase of the internal friction and elastic modulus of the fiber-reinforced soil. Having said that, the potential of fiber-reinforced soil in the engineering application, especially on the earth structure is really feasible.

Figure 5.5 illustrates an example of the randomly distributed fiber-reinforced soil. The fiber-reinforced soil also helps reduce the lateral earth pressure significantly, thereby decreasing the thickness of the wall, as the systematically reinforced soil backfill works. The mechanism of this system is explained in detail by Shukla (2017).

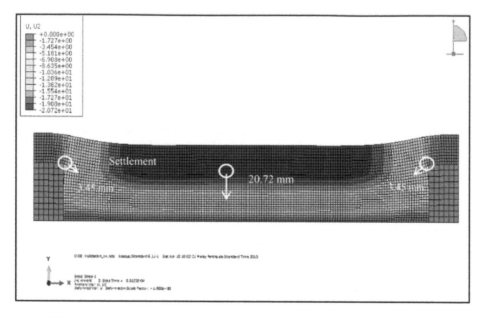

*Figure 5.4* Fiber reinforced soil as pile supported earth platform

Source: Anggraini et al. (2015)

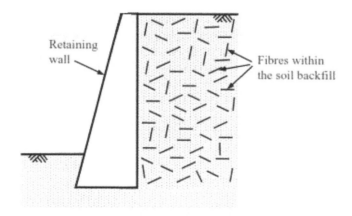

*Figure 5.5* A retaining wall supporting a randomly distributed fiber-reinforced soil backfill

Source: Shukla (2017)

## 5.3   Geosynthetics

Geosynthetics can be divided into three categories: permeable, referred to as geotextiles; impermeable fabric, or geomembranes; and geogrids. Geogrids are meshes with large openings either made from high-density polyethylene (HDPE) or knitted high-strength polyester coated with polyvinyl chloride (PVC). The permeable fabric can be further subdivided into

*Figure 5.6* Geosynthetics family
Source: Courtesy of MTS Fibromat Sdn Bhd

two types: woven and non-woven fabric. Their primary purpose is to act as reinforcement. Geosynthetics are thin, flexible, sheet-like materials enhancing the engineering performance of soils, as shown in Figure 5.6.

### 5.3.1   Geotextiles

The term "geotextile" is derived from "geo" and "textile" and may be simply defined as textile material used in a soil (geo) environment. The commonly used geotextiles today are either woven, non-woven or knitted geotextiles.

#### 5.3.1.1   Permeable woven geotextile

Figure 5.7 shows an example of an earlier version of permeable woven fabric. As shown, the fabric is made from woven filaments which are square in cross-section. Newer generation fabrics are made of filaments of various fibers or twisted tapes, as shown in Figure 5.8 for fabric woven geotextile and in Figure 5.9 for natural woven geotextile. The advantage of this newer generation woven fabric is that it is faster to weave and so its production cost is kept low.

Specific orientation of the fibers in a woven fabric means that the fibers share more or less equally in resisting an applied load in its load carrying direction (i.e., the warp direction of the fabric). Because of this, woven fabric usually has high modulus and low extension ability. Also, the woven fabric has anisotropic properties. Its tensile strength is normally higher in the warp (width of fabric) compared with the weft (length of fabric).

Therefore, these fabrics are often the choice in applications where tensile strength and high modulus in one direction is an important consideration compared with other directions. An example is in the construction of a road embankment. This type of fabric is also used in applications such as slope reinforcement and coastal erosion control.

Monofilament-on-monofilament

Monofilament-on-tape

Monofilament-on-multifilament

Multifilament-on-multifilament

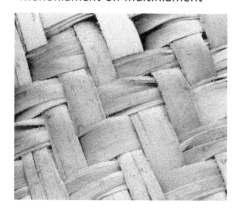

Slit film tape-on-slit film tape

Extruded tape-on-extruded tape

*Figure 5.7* Woven synthetic fabric geotextiles

Source: Courtesy of MTS Fibromat Sdn Bhd

Knitted base

Upper surface

*Figure 5.8* Woven natural geotextiles

Source: Courtesy of MTS Fibromat Sdn Bhd

Mechanically bonded nonwoven

Thermally bonded nonwoven

*Figure 5.9* Mechanically and thermally bonded non-woven, composed of fiber or filaments

Source: Courtesy of MTS Fibromat Sdn Bhd

### 5.3.1.2 Permeable non-woven synthetic fabric

Different from woven fabric, in a non-woven fabric the fibers are not oriented in a specific direction but at random. Several processes have been invented to produce these types of fabric for various purposes. Examples are spun bonded, needle punched and several others. Examples of these fabrics are shown in Figure 5.9 and Figure 5.10.

*Figure 5.10* Needle punched non-woven

Source: Courtesy of MTS Fibromat Sdn Bhd

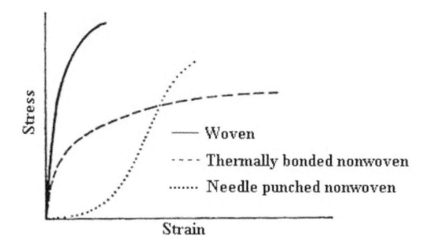

*Figure 5.11* Typical stress-strain relationship of a non-woven fabric compared with a woven fabric

Needle punched geotextiles are made by subjecting a web of fibers to the repeated entry of barbed needles that compact and entangle individual fibers to form a non-woven geotextile. Having said that, this method will provide a tighter geotextile and the fiber will not come off easily when doing the abrasion test. The main usage of the permeable non-woven fabric is as a separator and filter (drainage).

Due to its manufacturing process, the fibers of this fabric do not have a particular orientation. Therefore, theoretically its tensile strength is more or less the same in all directions. In general, this fabric is not as strong as woven fabric and has a higher extension ability value. Figure 5.11 shows a typical stress-strain relationship of a non-woven fabric compared with

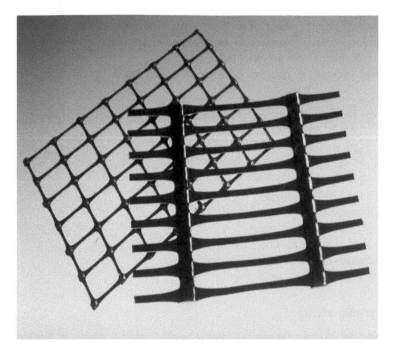

(a) Geogrid

(b) Geonet

*Figure 5.12* Netting

Source: Courtesy of MTS Fibromat Sdn Bhd

a woven fabric. Because the shape of the fiber is more or less uniform in the case of woven fabric, its pores in between are more or less the same size. This in contrast with the non-woven fabric, where the fibers are oriented at random.

The allowable strengths of geosynthetics are indicated below:

$$T_a = T_{ult} \frac{1}{RF_{ID} \times RF_{CR} \times RF_{CR} \times RF_{BD}} = T_{ult} \frac{1}{RF}$$

where

Ta = allowable tensile strength
Tult = ultimate tensile strength
$RF_{ID}$ = reduction factor for installation damage
$RF_{CR}$ = reduction factor for creep
$RF_{CD}$ = reduction factor for chemical degradation
$RF_{BD}$ = reduction factor for biological degradation
RF = overall reduction factor.

### 5.3.2   Geogrids/netting

Figure 5.12a and 5.12b show examples of nettings made from synthetic polymers. The main uses of these nettings are for soil stabilisation (also commonly known as geogrids, Figure 5.12a), growing grass, and slope erosion control (or geonet, Figure 5.12b).

#### 5.3.2.1   Impermeable fabric/geomembranes

There are two types impermeable fabrics that are available: (a) sheets of continuous extruded plastic materials and (b) fabric interwoven with nets laid on top and tied to impermeable plastic sheets. Among the main usages of these fabrics are for impermeable lining for a reservoir to prevent loss of water, impermeable lining for tunnels and for roof sealers as shown in Figure 5.13.

*Figure 5.13* Impermeable sheets acting as barriers to fluids

Source: Courtesy of MTS Fibromat Sdn Bhd

### 5.3.3   Geocomposite

Geocomposite is a non-woven geotextile, reinforced with a grid polyester yarns that are used in the making of high-strength geotextile. This combination provides the features of separation and filtration of the non-woven together with that of knitted high-strength yarn reinforcement (Figure 5.14).

Application of geocomposites include supporting and reinforce steep slopes, enhancing slope drainage, embankment reinforcement and stabilisation of landfill slopes, and as basal reinforcement for roads and storage areas.

### 5.3.4   Geocells

Geocells are essentially a series of strips of stiff polyethylene plastic of about 200 mm wide that are spot-welded together such that when stretched out they form a "honeycomb" type of arrangement. The honeycomb is laid on the soft ground and filled with granular backfill. The underlying principle is simple: granular material is stronger and stiffer when it is confined. During construction, it is also common to lay a light fabric (geotextile) beneath the geocells to act as a separator membrane. Geocells are usually referred to as a cellular confinement system (Figure 5.15)

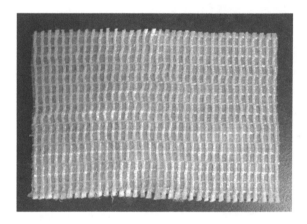

*Figure 5.14* Reinforcing geocomposite
Source: Courtesy of MTS Fibromat Sdn Bhd

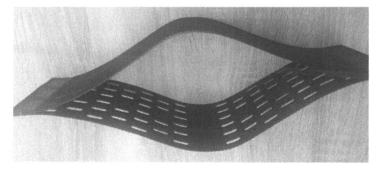

*Figure 5.15* Geocell
Source: Courtesy of MTS Fibromat Sdn Bhd

## 5.4 Geosynthetics usage in civil engineering

In the previous section we briefly listed the different types of fabrics that are available. In this section we will briefly describe the usage of the fabrics in civil engineering. Although there are many applications of synthetic fabrics available nowadays, their main functions can be categorised as follows (Table 5.2):

*Table 5.2* Functions of geosynthetics

| Function | Symbol | Description | Suitable products |
|----------|--------|-------------|-------------------|
| Filtration | | Allow the passage of fluids preventing the migration of soil particles | Geotextiles, geocomposites |
| Drainage | | Transport of fluids | Geonets, geocomposites |
| Separation | | Prevent the mixing of two different soils or materials | Geotextiles, geocomposites |
| Reinforcement | | Provide tensile forces in the soil mass, increase the bearing capacity, provide tensile and fatigue resistance | Geogrids, woven geotextiles, geocomposites |
| Erosion control or surficial stabilisation | | Avoid the detachment and transport of soil particles by rain, runoff and wind, root anchorage | Geomats, geocells |
| Impermeabilisation | | Fluid barrier | Geomembranes, geocomposites |

| Function | Symbol | Description | Suitable products |
|---|---|---|---|
| Confinement | 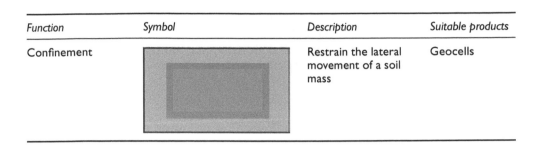 | Restrain the lateral movement of a soil mass | Geocells |

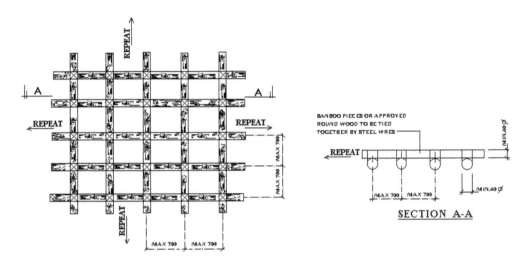

PLAN OF GEOTEXTILE FASCINE MATTRESS

*Figure 5.16* Non-woven fabric – bamboo mattress (Toh et al., 1994) railway – NW as separator

## 5.4.1   *Separation*

Fabric in this context is used to separate the different construction materials, for example:

- As a separator for zones of different materials in an embankment, earth dam or rock-fill dam.
- As a separator to separate the stone base from sub-grade beneath airfield runways, road pavement, car parks and temporary roads, to prevent occurrence of punching.
- As a separator for railway ballast from soil sub-grade to prevent punching.

Usually non-woven fabrics are used for these applications. These non-woven fabrics may also be used in combination with natural materials such as bamboo mattresses to aid in construction on very soft ground, such as on peat. An example of a case history on this is the construction of the Kuching Ring Road, north of Kuching city (Toh et al., 1994). In this case, bamboo (Figure 5.16) was used to aid in the rolling out of the non-woven fabrics over a very soft peat, in order to provide a separation layer on which a working platform was then developed.

The fabric acts as long-term separator for the various soils (subsoil and embankment fill), particularly in the case of a dynamic load, as illustrated in Figure 5.17 and Figure 5.18.

Combining Fibrocell with selected infill material will improve and enhance the load-bearing capacity by significantly reducing sub-grade contact pressure through lateral road distribution. This will retard the creation of deep grooves and prevent movement of soil particles.

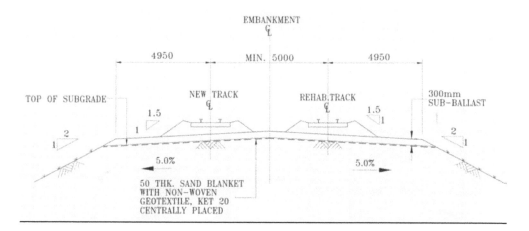

*Figure 5.17* Non-woven geotextile uses between the ballast and the subgrade (filter and separator)
Source: Courtesy of MTS Fibromat Sdn Bhd

*Figure 5.18* Subgrade project laying non-woven geotextile on the subgrade project Malaysia
Source: Courtesy of MTS Fibromat Sdn Bhd

## 5.4.2   Reinforcement

In cases of construction involving soils with low bearing capacity, say for construction of a road embankment on very soft ground, a problem may arise with the short-term stability of the subsoil. In this case fabric can be used mainly as reinforcement to assist the construction. The bearing capacity of the subsoil can be improved too (Figure 5.19).

A construction of road on soft soil using synthetic fabric is shown in Figure 5.20. Synthetic fabrics are generally available in roll form, of width 3–5 m by more than 100 m long. These fabrics are then cut according to work requirements. There are two types of jointing systems that are normally used: sewing and overlap. Typical overlap width is 0.5–1.0 m. The sewing method, however, is generally preferred as it is more economical. It must be noted that all joints must be made along the warp direction of the fabric.

Woven fabrics, unlike non-woven fabrics, generally have higher strength in the warp direction (i.e., the load-bearing direction of the fabric) than in the weft direction. Therefore, in the case of simple reinforcement, say an embankment on soft ground, fabric sheets are laid in such a way that the warp direction of the fabric will coincide with the width of the embankment. However, in special cases where plane strain condition no longer exists, at least two layers of fabric are laid with the warp and weft direction of the fabric to overlap at 90° angles with each other.

In most cases of subsoil reinforcement, fabric sheets are often laid direct on the soft ground. Only minimal site clearing will be done. However sometimes a layer of sand is first laid for the purpose of drainage. Soft soils usually exist in low-lying and waterlogged areas. In such cases the reinforcing fabrics are often laid directly above the water.

*Figure 5.19* Construction of road embankment at Batu Kawan, Seberang Perai, Penang

Source: Courtesy of MTS Fibromat Sdn Bhd

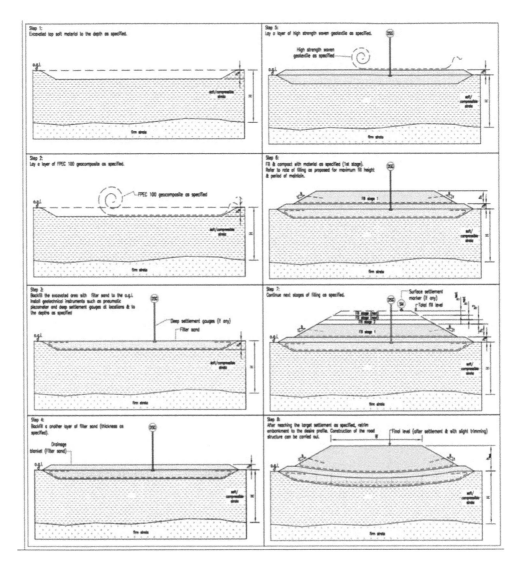

*Figure 5.20* Guidelines for construction of high-strength woven geotextile fill road embankment

Source: Courtesy of MTS Fibromat Sdn Bhd

Wall reinforcement is a vertical and steep slopes can be contracted by reinforcing the soil with geogrids and reinforcement geotextiles in Figure 5.21 and Figure 5.22.

Usually the woven fabrics or geogrids are used for reinforcement purposes. Sometimes these fabrics are also used in special constructions, such as in the piled embankment (see Figure 5.23).

Augustin and Al-Obaidy (2003) describe the special use of fabric reinforcement for construction of a new railway embankment and track work on soft ground. Figure 5.23 shows a model of the embankment cross-section.

Slopes encountering difficulties in vegetative growth can make full use of Fibrocell. Fibrocell with fertilised infill soil material will ensure vegetative growth and hence prevent surface erosion for acidic or hard rocky steep slopes (Figure 5.24).

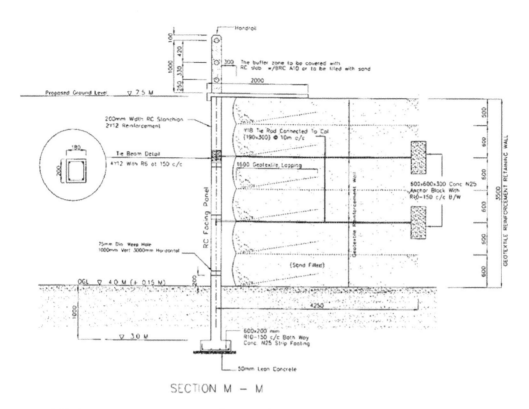

SECTION M – M

*Figure 5.21* Cross-section construction of a steep slope

Source: Courtesy of MTS Fibromat Sdn Bhd

*Figure 5.22* Reinforcement of wall of a steep slope

Source: Courtesy of MTS Fibromat Sdn Bhd

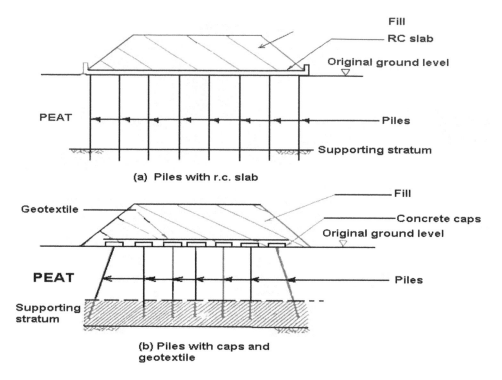

(a) Piles with r.c. slab

(b) Piles with caps and geotextile

*Figure 5.23* Fabric application in piled embankment

*Figure 5.24* Geocell as slope protection

Source: Courtesy of MTS Fibromat Sdn Bhd

*Figure 5.25* Geocell as retaining wall system
Source: Courtesy of MTS Fibromat Sdn Bhd

Geocell can be used to construct flexible retaining wall system to optimise space usage. When stacked with a setback, Fibrocell will form a stable gravity retaining wall system. The infill material can be either granular or concrete, depending on the function of the infill material and its design (Figure 5.25).

### 5.4.3 *Drainage*

Although there are many types of drains and filter designs available, the drainage system as shown in Figure 5.26 is perhaps the simplest and most commonly known. The main use of fabric in drainage can be described as follows:

• To prevent migration of soil fines into the crushed stone filter or underground drain system. Fabric is used to replace the need for a graded filter system comprising layers of various sizes of sand and gravel.
• To prevent penetration and loss of coarse material with high permeability.

Figure 5.26  Using non-woven geotextile as subsoil drainage system
Source: Courtesy of MTS Fibromat Sdn Bhd

### 5.4.4  Erosion control

Typical situations where synthetic fabric is used for erosion control are:

- Coastal erosion control, where fabric is used to prevent erosion of the in situ soil and allow weeds to grow
- As riprap or gabion for coastal control, as shown in Figure 5.27
- As control for erosion at sewer spillways.

Geocell can be used to protect channels or riverbanks against erosion by strong and rapid water flows. Geocell is effective at preventing scouring and protecting shallow slope failures. It is also an economic solution to solving problems relating to irrigation and drainage (Figure 5.28).

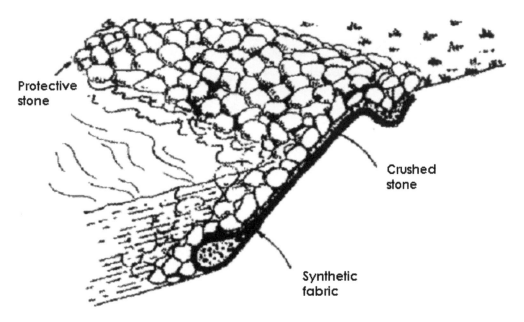

*Figure 5.27* Fabric for coastal protection

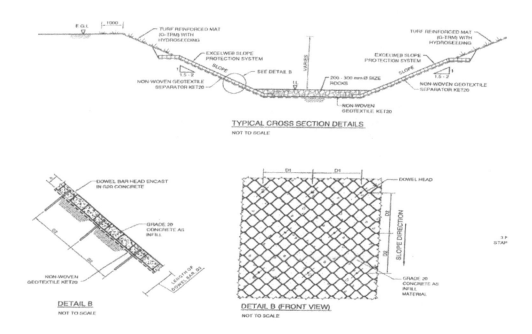

*Figure 5.28* Geocell as channel protection
Source: Courtesy of MTS Fibromat Sdn Bhd

*Figure 5.29* Geosynthetic clay liners at municipal solid waste project, Malaysia

Source: Courtesy of MTS Fibromat Sdn Bhd

### 5.4.5  Liners

Impermeable fabrics are used as liners for water reservoirs to reduce loss of water and also to line tanks and tunnels. Geosynthetic clay liners (GCLs) are geotextile and bentonite composites (typically sodium bentonite sandwiched between two layers of geotextile) engineered for a variety of environmental containment applications. Figure 5.29 shows the laying of geosynthetic clay liners at a municipal solid waste project.

## 5.5  Use of synthetic reinforcing fabric for construction of embankments on soft ground

In the previous section we saw many uses of synthetic fabrics in civil engineering constructions. Perhaps the most simple and yet effective use of fabric is in the construction of an embankment on soft ground, such as for roads, railways and runways.

In constructing an embankment on soft ground or soil with low bearing capacity, problems may arise from the short-term instability of the embankment and loss of bearing capacity of the founding soil. However, there are a number of methods that could be used to solve this problem:

- Removal of the upper soft layers until reaching a stratum with sufficient bearing capacity and then backfilling the excavation with good quality fill material.
- Stage construction: placing earth fill in a sequence of thin layers with a long rest period in between each of the layers. This is to allow excess pore water pressure to dissipate and hence increase in the shear strength of the subsoil.

However, the alternative use of synthetic fabrics in this case will give a number of advantages:

- Lateral displacement and loss of short-term equilibrium can be controlled. Safety factors against overall stability can be improved.

- Differential settlement due to local shear failure can be reduced.
- A thicker layer of earth fill can be placed in a single operation without loss of embankment stability.
- The width of the embankment (toe to toe) to achieve the design height can be reduced. With this, the quantity of earth fill required to build the embankment can also be reduced.

## 5.6    Factors influencing design of embankment reinforced with fabrics

The concept of using reinforcing fabric to reinforced soil can be considered theoretically sound. This has been proven in laboratory studies. For example, Broms (1977) performed a series of triaxial tests with fabric buried in loose and dense sand. According to Broms, the failure stress of the reinforced sand increased due to the presence of the fabric. Andrawes et al. (1983), who had conducted studies on model footing to study behavior of sand reinforced with fabric, concluded that the reinforcement increased the bearing capacity of the sand.

The main objective in designing an embankment reinforced with fabrics is to place the fabrics so that they act as an effective tensile reinforcement. Therefore the fabric has to be placed in the tensile zone and oriented in the direction of major strain. Fabrics buried in soil in the tensile zone will anisotropically restrain the natural expansion of the soil. However, if the fabric is close to the line of zero extension, then theoretically slippage can occur at the interface between the fabric and the soil. These conclusions are derived from an extensive study on the stress strain characteristics of soil by Roscoe (1970).

Reinforcing fabrics can generally be placed in the horizontal direction only. Nevertheless a horizontal fabric layer will be correctly oriented in the main body of the embankment. But this is not the case under the embankment slope, especially near the embankment toe. This is because as stated earlier, reinforcement outside the tensile zone will encourage rather than prevent failure because the friction or adhesion at the soil-fabric interface is less than the friction or adhesion of the soil alone. However, the reinforcing fabrics can temporarily be extended to the slope face and then back folded to cover the slope face. The free ends of the fabrics can be anchored to the soil using suitable anchors. With this technique, the soil close to the slope face will be strengthened by normal stresses exerted by the fabrics near the slope face.

In the previous section, we saw various types of fabrics that are currently available. The fabrics are not only made with various processes (such as woven, non-woven or extruded) but also with different organic polymers (such as polyester, polypropylene, polyethylene and nylon). These fabrics clearly have variable properties.

In general, woven fabric functions better as a reinforcement compared with non-woven fabric. This is because in woven fabric, its fibers share more or less equally in resisting a load. Because of this, the modulus of a woven fabric is high and its extensionability is low.

Synthetic fabrics can also lose their strength due to creep and environmental degradation, exposure to ultraviolet light, chemical and biological attack and prolonged exposure to heat. The resistance of synthetic materials, however, varies from one to another.

Synthetic fabric can also degrade due to environmental attack. For example, polyester will rot if left too long submerged in water (Rankilor, 1981), while polypropylene shows a rather good resistance to microbiological attack.

Soil generally contains both organic and inorganic chemicals. Inorganic chemicals are generally products of rock and mineral weathering while organic chemicals are from decayed plants and animal carcasses. Compared with other polymers, polyester shows a better resistance to chemical attack. Polyamide, on the other hand, will degrade rapidly when buried too long within soil with high acidity.

In general, thermal degradation is not an important consideration in most engineering construction. For example polyester will only soften at temperatures higher than about 200°C, while polypropylene and polyethylene will soften at temperatures between 120°C to 160°C.

From the preceding facts, it can be seen that polymers and the manufacturing processes of particular fabrics play an important role in determine the characteristics of the manufactured fabric. For example, for use as reinforcement, characteristics such as stress-strain, creep and environmental degradation have to be considered. Reinforcing fabric has to have high modulus with low creep and extensibility coefficient. It has also been seen that all fabrics degrade due to environmental attack, like exposure to ultraviolet light and chemical and biological attack. Only the rate of degradation differs, according to the type of polymers and additives used. Perhaps the simplest and yet the most effective use of reinforcing synthetic fabric is in the construction of an embankment on soft ground, where the critical period is at the end of the construction. With time, excess pore water pressure generated by the embankment construction will dissipate, and the subsoil will gain shear strength through the process of consolidation. With this, the loss of fabric strength due to the environmental degradation will be replaced by the gain in soil shear strength.

## 5.7   Design method

Calculation for embankment stability can be made using the following methods:

- Finite element method
- Limit equilibrium method.

Theoretically, the finite element method is a powerful method. It can take into account material non-linearity and variable geometries. But this method needs to be performed with the aid of a computer.

The limit equilibrium method, although theoretically not the best method, is easy to use in practice. The main criticism of this method is that its accuracy is very dependent on the assumed failure. Mohr-Coulomb criteria are assumed to operate along a postulated slip surface. In general, this method provides information during soil failure, but do show soil behavior prior to the collapse.

In the previous section we have described the modified limit equilibrium method to calculate embankment stability (i.e., for case of subsoil reinforcement). There are four modes of failure that need to be taken into account in the analysis. Common solutions used to solve stability and settlement problems for embankments on soft ground can be seen in Figure 5.34.

### 5.7.1   Bearing capacity failure

A method to determine the bearing capacity of soil was derived by Prandtl in 1951. Bearing capacity is first checked to ensure it is sufficient to carry the embankment load. If not, the embankment may have to be built in stages.

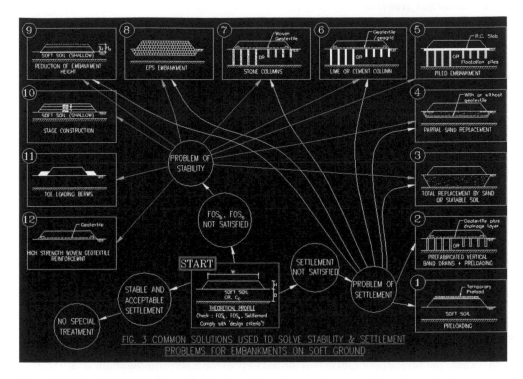

*Figure 5.30* Common solutions used to solve stability and settlement problems for embankments on soft ground

Source: Courtesy of MTS Fibromat Sdn Bhd

Equation for bearing capacity for strip load:

$$q_f = c \, N_c + q \, N_q + \tfrac{1}{2} \gamma \, B. \, N_g \tag{5.1}$$

where $N_c$, $N_q$ and $N_\gamma$ are bearing capacity factors.

$$N_c = \exp\left(\pi \tan \phi\right) \tan 2 \left(45 + \phi/2\right) \tag{5.2}$$

$$N_q = \left(N_q - 1\right) \cot \phi \tag{5.3}$$

$$N_\gamma = 1.5 \left(N_q - 1\right) \tan \phi \tag{5.4}$$

where

$c$ = cohesion
$\phi$ = angle of friction
$\gamma$ = effective unit weight of soil
$B$ = footing width.

### 5.7.2   Internal stability

For embankments built on saturated clay soil, the undrained shear strength of the soil is normally used in the stability calculation. It is assumed that the in situ upper soil layer is not sufficient to restrain the active pressure from the embankment fill. The embankment therefore has to deform in the horizontal direction.

From Rankine's theory, active pressure is:

$$Pa_1 = \frac{1}{2} \cdot K_a \cdot \gamma \, H^2 \tag{5.5}$$

The restraining force of the fabric required is:

$$S_1 = F \cdot Pa_1 \tag{5.6}$$

where

$K_a$ = Rankine's active pressure coefficient = $(1 - \sin \phi)/(1 + \sin \phi)$
$\gamma$ = soil unit weight (kN/m²)
$\phi$ = friction angle
$F$ = factor of safety.

### 5.7.3   Foundation stability

In this case, the reinforcing fabric is required to provide a force to maintain equilibrium of a foundation soil block. A tension crack is assumed to form through the entire height of the embankment. With this the embankment is assumed only to function as overburden on the subsoil. Reinforcement force is calculated using bonding stress between the foundation soil and reinforcement only. No bonding stress is assumed to be mobilised between the reinforcement and the embankment. This is because the foundation soil block is only restrained by shear force at the lower part of the reinforcement during failure.

Assuming soil to be fully saturated, $\phi_u = 0$. Therefore, for effective stress analysis:

$$Pa_2 = \frac{1}{2} \gamma \, h^2 - 2c_u \cdot h + qs_1 \cdot h \tag{5.7}$$

$$P_p = \frac{1}{2} \gamma \, h^2 + 2c_u \cdot h + qs_2 \cdot h \tag{5.8}$$

For equilibrium:

$$P_p + c_u \cdot L + S_2 \le Pa_2 \tag{5.9}$$

Therefore:

$$S_2 = F \cdot Pa_2 - P_p - c_u \cdot L \tag{5.10}$$

where $Pa_2$ = active force, $P_p$ = passive force, $\gamma$ = soil unit weight, h = soil block height (assumed), $c_u$ = undrained cohesion, qs1 and qs2 = surcharge and H = embankment height.

Although the reinforcing force required for foundation stability is in general smaller than that for the overall stability, the criteria of foundation stability usually determine the minimum acceptable width of the embankment.

### 5.7.4   Overall stability

Using the Ordinary or Bishop method of analysis, the embankment is first check for overall stability for case without reinforcement. The critical slip surface is found using a trial and error method.

$$\text{Factor of safety (un reinforced)} = \text{restraining moment/disturbing moment}$$

$$= \text{MR/MD} \qquad (5.11)$$

$$= \left(\sum \tau_s . \delta_s\right) \text{R/W.x}$$

Using a critical slip surface found in the above unreinforced condition, further calculations can be made incorporating a number of assumptions as follows:

*   Reinforcement acts in the direction where it is placed
*   Reinforcement gives additional restraining moment, $\Delta$MR, where $\Delta$MR = P * y, P is reinforcing tensile force mobilised, and y is the lever arm
*   Factor of safety (improved) for embankment stability, F* = (MR + $\Delta$MR) /M.

Therefore force in fabric required for overall stability is:

$$S_3 = F * \times P \qquad (5.12)$$

### 5.7.5   Design forces

The maximum restraining force that can be obtained from the fabric, besides its ultimate tensile strength, is limited by the shear resistance (stress bond) between the reinforcement and subsoil. Therefore if the applied force is larger than the mobilised soil stress bond, the reinforcing fabric will "pull out" and the embankment will fail. Three important factors that influence the tensile reinforcement are:

*   Mobilised soil/reinforcement bond
*   Distribution of tensile strain inside the soil adjacent to and in direction of reinforcement
*   Load characteristic and reinforcement extensibility with time in the ground.

The available reinforcement force profile is determined for each of the limit conditions as follows:

1.  Choose the value of the soil stress bond with reinforcement on each side and at each point along the reinforcement
2.  Choose the tensile strain value in the soil in the direction of the reinforcement at each point along the reinforcement
3.  Determine the stress-strain relation of the reinforcement inside the ground by taking into account the time effect during which the reinforcement must act

4.  Maximum force can be calculated as follows:

    a.  Maximum force available is calculated from both ends of the reinforcement at rate given by the bonding stress as in (1) above.
    b.  Overall maximum force that can be mobilised is limited to the magnitude of strain allowable in the soil (2), corresponding to the reinforcement force as define by (3).

A factor of safety is applied to account for loss of strength in the fabric due to factors such as chemical attack, biological attack, degradation due to ultraviolet light and creep. A typical value for the factor of safety is 2, which means the available load is 50% of the fabric ultimate load. The rate of extensibility that can be accepted for a reinforcing fabric is generally about 5%–6%.

## 5.8  Case study

The case study is the South West District Penang. The total slope height is 6 m and the slope gradient is about 70° (Figure 5.31). A section of the proposed slope rehabilitation was modelled and checked for stability (Figure 5.32).

The acceptable slope stability safety factor (FoS) for the slope treatment is set to a minimum of 1.4. Using the Reinforced Slope Stability Analysis (ReSSA) (Figure 5.36), numerous trial slip circles were tried out and the slip with the lowest safety factor was presented. All the assumptions need to be verified first before the construction of this

*Figure 5.31* Slope stabilisation work

Source: Courtesy of MTS Fibromat Sdn Bhd

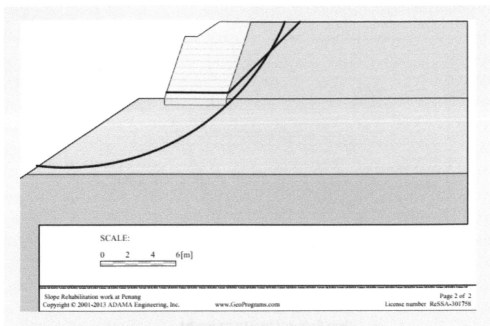

SCALE:

0    2    4    6[m]

*Figure 5.32* Critical result of rotational and translational stability analyses

Source: Courtesy of MTS Fibromat Sdn Bhd

*Table 5.3* Soil engineering parameters

| Layer (top down) | Description | Soil Model | Bulk unit weight, $\gamma$ (kN/m³) | Design cohesion, $C'$ (kN/m²) | Design soil internal friction, $\phi$ (°) |
|---|---|---|---|---|---|
| 1 | Backfill soil or Reinforced soil | Mohr Coulomb | 18 | 5 | 29 |
| 2 | Existing/Retained soil | Mohr Coulomb | 18 | 10 | 30 |
| 3 | Foundation soil –crusher | Mohr Coulomb | 18 | 0 | 36 |
| 4 | Run fill base Foundation soil 1 | Mohr Coulomb | 18 | 10 | 32 |

slope stabilisation works. Typical section of reinforced fibrogrid slope wall is presented in Figure 5.37. The related assumed design parameters are shown in Table 5.3.

The allowable long-term design tensile for the reinforcement FG100, for example, is:

$$Ta = Tu/(f1 + f2 + f3) = 47 \text{ kN/m}$$

Both internal and global stabilities were checked and the lowest safety factors were determined. The result of the analyses showed satisfactory safety factor using different wall heights with Reinforcing fibrogrid elements FG80 as shown in the drawing.

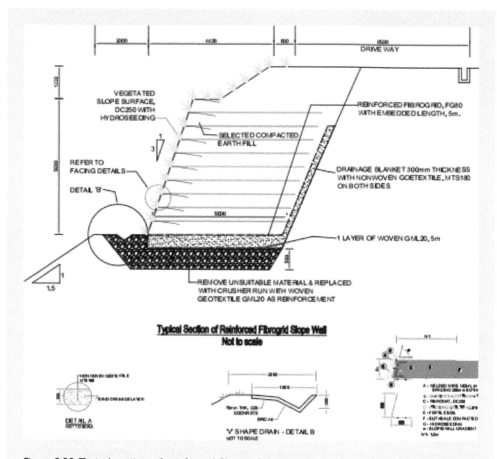

Figure 5.33 Typical section of reinforced fibrogrid slope wall

Source: Courtesy of MTS Fibromat Sdn Bhd

Table 5.4 Fibrogrid parameters

| Ultimate tensile, Tu (kN/m) | Installation damage factor, f1 | Durability factor, f2 | Creep factor, f3 | Design value, T des (kN/m) |
|---|---|---|---|---|
| FG80–80 | 1.05 | 1.1 | 1.47 | 47 |
| GML20–200 | 1.02 | 1.10 | 1.42 | 125 |

Table 5.5 The different FOS for the wall height

| Wall height, H (m) | FOS (Global stability) | FOS (Internal stability) | FOS (General wedge) |
|---|---|---|---|
| 5.0 | 1.44 | 2.62 | 1.73 |

Basically, the design is a combination of the following elements:

1. Drainage controls: subsoil drainage media to cater for the underground water and infiltration during and after construction.
2. Slope reinforcement: the use of fibrogrid and geogrid as reinforcement behind the slope/wall system not only stabilised the slope but also encourages grass to grow and became an environment "greening effect," especially in rural and urban areas.
3. Temporary protection: the excavation for the wall construction will ensure the short-term stability of the excavation slopes.

# References

Abtahi, S. M., Sheikhzadeh, M. & Hejazi, S. M. (2010) Fiber-reinforced asphalt-concrete–a review. *Construction and Building Materials*, 24(6), 871–877.

Andrawes, K. Z., McGown, A. & Wilson-Fahmy, R. F. (1983) The behaviour of a geotextile reinforced sand loaded by a strip footing. *Proceedings of 8th European Conference on Soil Mechanics and Foundation Engineering*, Helsinki, Balkema, Rotterdam.

Anggraini, V., Asadi, A., Farzadnia, N., Jahangirian, H. & Huat, B.B.K. (2016a) Effects of coir fibres modified with Ca (OH) 2 and Mg (OH) 2 nanoparticles on mechanical properties of lime-treated marine clay. *Geosynthetics International*, 23(3), 206–218.

Anggraini, V., Asadi, A., Farzadnia, N., Jahangirian, H. & Huat, B.B.K. (2016b) Reinforcement benefits of nanomodified coir fiber in lime-treated marine clay. *Journal of Materials in Civil Engineering*, 28(6), 06016005.

Anggraini, V., Asadi, A., Huat, B. B. & Nahazanan, H. (2015) Performance of chemically treated natural fibres and lime in soft soil for the utilisation as pile-supported earth platform. *International Journal of Geosynthetics and Ground Engineering*, 1(3), 28.

Anggraini, V., Asadi, A., Syamsir, A. & Huat, B. B. (2017) Three point bending flexural strength of cement treated tropical marine soil reinforced by lime treated natural fiber. *Measurement*, 111, 158–166.

Augustion, P. C. & Al-Obaidy, J. H. (2003) Construction of new railway embankment and trackworks on soft ground adjacent to an existing mainline track at Pondok Tanjung for KTMB. In: Huat et al. (eds) *Proceedings of 2nd International Conference on Advances in Soft Soil Engineering and Technology, July 2003*. Universiti Putra Malaysia Press, Putrajaya. pp. 635–648.

Bell, F. G. (1996) Lime stabilization of clay minerals and soils. *Engineering geology*, 42(4), 223–237.

Broms, B. B. (1977) Polyester fabric as reinforcement in soils. *Conference on Use of Geotextile in Geotechnique*. Paris (n.p.).

Consoli, N. C., Montardo, J. P., Donato, M. & Prietto, P. D. (2004) Effect of material properties on the behaviour of sand–cement–fibre composites. *Proceedings of the Institution of Civil Engineers-Ground Improvement*, 8(2), 77–90.

Freitag, D. R. (1986) Soil randomly reinforced with fibers. *Journal of Geotechnical Engineering*, 112(8), 823–826.

Huat, B.B.K. & Aziz, A. A. (1988, February) Penggunaan Fabrik Sintesis dalam Pembinaan Benteng Jalan Raya. *Bulletin Institution Engineers Malaysia*, 2, 21–24.

Jenck, O., Dias, D. & Kastner, R. (2005) Soft ground improvement by vertical rigid piles: Two-dimensional physical modelling and comparison with current design methods. *Soils and Foundations*, 45(6), 15–30.

Kaniraj, S. R. & Gayathri, V. (2003) Geotechnical behavior of fly ash mixed with randomly oriented fiber inclusions. *Geotextiles and Geomembranes*, 21(3), 123–149.

Lekha, B. M., Goutham, S. & Shankar, A.U.R. (2015) Evaluation of lateritic soil stabilized with Are-canut coir for low volume pavements. *Transportation Geotechnics*, 2, 20–29.

Lin, T., Jia, D., Wang, M., He, P. & Liang, D. (2010) In situ crack growth observation and fracture behavior of short carbon fibre reinforced geopolymer matrix composites. *Materials Science and Engineering A*, 527(9), 2404–2407.

Maher, M. H. & Ho, Y. C. (1994) Mechanical properties of kaolinite/fiber soil composite. *Journal of Geotechnical Engineering*, 120(8), 1381–1393.

Marandi, S. M., Bagheripour, M. H., Rahgozar, R. & Zare, H. (2008) Strength and ductility of randomly distributed palm fibers reinforced silty-sand soils. *American Journal of Applied Sciences*, 5(3), 209–220.

Mirzababaei, M., Mohamed, M. H., Arulrajah, A., Horpibulsuk, S. & Anggraini, V. (2018) Practical approach to predict the shear strength of fibre-reinforced clay. *Geosynthetics International*, 25(1), 50–66.

Musenda, C. (1999) *Effects of fiber reinforcement on strength and volume change behavior of expansive soil*. Doctoral dissertation, MS Thesis, The University of Texas at Arlington, Arlington, TX.

Park, S. S. (2011) Unconfined compressive strength and ductility of fiber-reinforced cemented sand. *Construction and building materials*, 25(2), 1134–1138.

Park, T. & Tan, S. A. (2005) Enhanced performance of reinforced soil walls by the inclusion of short fiber. *Geotextiles and Geomembranes*, 23(4), 348–361.

Prabakar, J. & Sridhar, R. S. (2002) Effect of random inclusion of sisal fibre on strength behaviour of soil. *Construction and Building Materials*, 16(2), 123–131.

Ramaswamy, S. D. & Aziz, M. A. (1982) Jute fabric in road construction. *2nd International Conference on Geotextile*. Industrial Fabrics Association International, Nevada.

Rankilor, P. R. (1981) *Membranes in Ground Engineering*. John Wiley & Sons, New York.

Roscoe, K. H. (1970) The influence of strains in soil mechanics. *10th Rankine Lecture, Geotechnique*, 20(2), 129–170.

Shukla, S. K. (2017) *Fundamentals of Fibre-Reinforced Soil Engineering*. Springer, Singapore.

Sivakumar Babu, G. L., Vasudevan, A. K. & Sayida, M. K. (2008) Use of coir fibers for improving the engineering properties of expansive soils. *Journal of Natural Fibers*, 5(1), 61–75.

Toh, C. T., Chee, S. K., Lee, C. H. & Wee, S. H. (1994) Geotextile-bamboo fascine mattresses for filling over very soft soils in Malaysia. *Geotextile and Geomembranes*, 13, 357–369.

# Chapter 6

# Shallow stabilisation

## 6.1 Introduction

Existing soil on site may not be suitable to support structures such as buildings, roads or embankments. In its simplest form, soils on site are at least compacted to their maximum dry densities, thus improving their shear strength. In cases where in situ soils are found unsuitable, they need to be removed and replaced with better quality imported soils. This imported soil, however, also needs to be properly compacted to enable it to support the structure's load.

Occasionally the properties of the in situ soils can be improved by adding suitable stabilising agents. In this chapter we will discuss methods of shallow stabilisation using chemical additives such as cement, lime and other admixtures. Lately there has been an increase in using blended binders. Mixing cement with slag, fly ash and polymers has gained in popularity. The prime propose of blending binders is to reduce costs, improve technical performance and provide relatively long-term durability. This chapter deals with single and blended binders in different soils and with the techniques to install and evaluate the effects of blending binders. Such chemicals may also be used to stabilize thick deposits of soil, especially soft soil. This is better known as deep stabilisation, and will be discussed in further detail in Chapter 7 of this book.

## 6.2 General principles of additive stabilisations

As mentioned earlier, sometimes the properties of the in situ soil can be improved by adding suitable chemical additives. The most commonly used chemicals are lime and cement. The main objectives of stabilisation are:

- To modify the soil properties
- To speed up construction
- To improve strength and durability of the soil.

## 6.3 Types of additives used in soil stabilisation

### 6.3.1 Lime stabilisation

Lime is normally used to stabilise fine-grained soils. Lime is produced by calcination of limestone or dolomite at high temperatures (about 900°C). Types of lime available are:

1. Hydrated lime ($Ca(OH)_2$)
2. Quicklime (CaO)

3.    Monohydrated dolomite lime $(Ca(OH)_2 + MgO)$
4.    Dolomite quicklime.

The amount of lime normally used to stabilise most types of soil ranges between 5% and 10%.

Quicklime is more efficient to effect changes in soil strength compared with hydrated lime, but quicklime is quite dangerous as it can destroy live tissue.

When lime is added to soil, a number of chemical reactions take place:

1.    Exchange of cations
2.    Flocculation and aggregation
3.    Pozzolanic reaction.

The cation exchange reaction and flocculation-aggregation result in changes in the clay texture, whereby the clay platelets will combine to form larger particles. Due to this reaction, the liquid limit of the soil will be reduced while the plastic limit will be increased. As a result, the soil plasticity index will be reduced and the shrinkage limit will be increased. Therefore the workability of the soil will be enhanced and the soil strength, engineering and deformation properties will be improved.

Some of the effects of lime stabilisation are shown in Figures 6.1 and 6.2. The pozzolanic reaction between soil and lime involves the reaction between lime with the soil silica and alumina to form a cementing material. This pozzolanic reaction may continue over a long period of time. High temperatures, however, speed up the strength increase of a lime-soil mixture. The lime stabilisation is therefore particularly suitable in areas with a hot climate. The moisture content at which the soil is mixed with lime can also be critical since the optimum moisture content for maximum shear strength need not be the same for the maximum dry density. For clayey soil, the optimum moisture content for strength may be on the dry side of the optimum moisture content for maximum dry density, while for silty soil it may be on the wet side of the optimum. Figure 6.1 shows the effect of lime admixture on the compaction characteristics of the soil. In this case, the lime has an influence to reduce the maximum dry density of the compacted soil and increase its optimum moisture content.

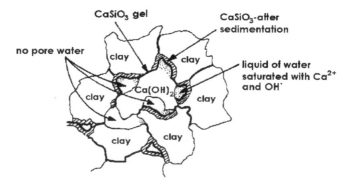

*Figure 6.1* Principle of soil stability with lime
Source: Van Impe (1989)

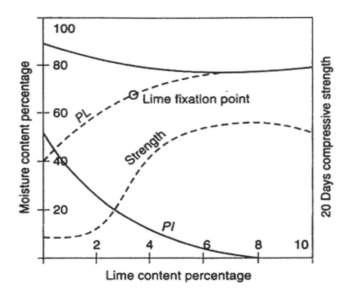

*Figure 6.2* Effect of lime content on the properties of clay-lime

Source: www.soilmanagementindia.com

Generally, an increase in plastic limit is first rapid and the rate decreases beyond a certain lime content (Figure 6.2). This point is often termed the lime fixation point. This is the approximate lime content that is considered to be used up for modification of clay. During this range, the increase in stability of the clay-lime may not be noteworthy. When the lime content in the mix is further increased, there is a high rate of increase in stability.

However, when the lime content is increased beyond a certain proportion, the stability values generally start decreasing. With proper lime treatment it is possible to make the clay almost non-plastic with the plasticity index reducing to practically zero.

An increase in lime content also causes considerable reduction in swelling and an increase in the shrinkage limit. All these changes are desirable for the stabilisation of clay.

### 6.3.1.1   Construction procedures

Lime stabilisation can be carried out using the following three methods.

- In situ or imported soil can first be broken down with a special mixer. Lime is then spread on the soil surface in a dry condition. Water is then added in a number of stages, with the soil and lime being mixed at each of the stages. When the final (required) water content has been added, the soil-lime mixture is then finally compacted with a roller.
- Soil with suitable water content and lime can be premixed at a special mixing plant, and then transported back to the site to be compacted.
- Lime slurry can be injected into the soil to depth of 2–3.5 m and spaced over a distance of 2–2.5 m. This technique is particularly suitable to prevent swelling of expansive soil. The technique also requires a layer of soil of about 0.25 m thick to be compacted using ordinary compaction methods.

Soil stabilised with lime needs to be cured in the same manner as curing a soil-cement mixture. The hardening process of lime, however, is much slower compared with cement. Therefore, soil stabilised with lime need not be compacted as soon as soil stabilised with cement. In other words, the working time with lime is not as critical as with cement stabilisation.

### 6.3.2    Cement stabilisation

The use of cement for the purpose of stabilisation began in 1917, when Dr. J. H. Amies patented a mixture of soil and cement, which he named soil amines.

Cement is used as a soil stabilising agent especially for road construction, such as for subbase, airport runways and earth dams. It is also used for the construction of low-cost houses, especially in arid regions. This material can be used to stabilise sandy and clayey soils. As with lime, cement has the effect to reduce the liquid limit and increase the plasticity index, hence increasing the workability of soil.

There are a number of factors that influence the soil-cement mixture:

*   Type and properties of soil
*   Quantity and type of cement
*   Soil moisture content
*   Mixing and compaction method
*   Condition and curing time.

In theory, any soil can be stabilised with cement. But increase in the silt and clay content requires more cement to be added. Soils most suitable to be stabilised with cement are a mixture of sand and gravel of good grade, and with less than 10% fines passing a 75 μm sieve, and with a coefficient of uniformity of not less than 5. The presence of organic and sulfate materials inside the soil may prevent the cement from hardening, therefore such soils may not be suitable. For clayey soil, cement stabilisation is effective when the soil fines (passing a 75 μm sieve) are less than 40%, the liquid limit is less than 45%–50% and the plasticity index is less than about 25. From this information, it is clear that granular soil and clayey soil of low plasticity are most suitable for stabilisation with cement compared with lime.

Any type of cement can be used to stabilise soil, but the most commonly used is ordinary Portland cement. The amount of cement normally used ranges from 6% to 14%.

As with lime, cement helps to increase soil strength. Strength increases with curing time, and high temperatures accelerate the process of strength gain. Therefore, as with lime stabilisation, cement stabilisation is suitable for areas with hot climates.

### 6.3.2.1    Construction procedures

Soil is first broken down with a special mixer. Cement is then spread on the soil surface in a dry condition. Water is then added in a number of stages, with cement being mixed with the soil at each of the stages. When the final (desired) water content has been added, the soil-cement mix is then finally compacted with a roller.

Soil and cement can also be premixed a special mixing plant and then transported to the site to be compacted.

After the mixing and compaction process is done, the stabilised soil is cured with one of the following methods:

- Covered with a layer of impermeable plastic
- Covered with a layer of soil of about 76 mm thick and intermittently sprayed with water to maintain a moist condition.

After curing, these materials are removed.

### 6.3.2.2   Other stabilisers

Besides lime and cement, there are also several other stabilisers that could be used to stabilise soils, including lime kiln dust, fly ash, ionic stabilisers and enzymes.

### 6.3.3   Fly ash

Nikraz et al. (2003) reported studies carried out on the use of a lime kiln dust and fly ash mixture for stabilising fine-grained soils. Lime kiln dust is a by-product in the process of creating quicklime. Lime kiln dust contains a significantly lower percentage of calcium than quicklime or hydrated lime and as such is cheaper.

Fly ash is a pozzolanic material which contains finely divided particles of siliceous and/or aluminous glass with small quantities of compounds like ferric and calcium oxide with minor amounts of crystalline constituents such as mullite, quartz, magnetite and hematite. Particle size is no larger than 250 μm in diameter and exhibits high mechanical strength, with a density range of 0.6–3 tonnes/m$^3$. In presence of lime and at higher pH, the particles of silica and alumina react with lime in the presence of water to form a gel of calcium silicate hydrate and calcium aluminate hydrate, in a manner similar to Portland cement hydrates. Normally, with the addition of water, the mixture of lime and fly ash reacts to form a new material with strength and durability properties superior to those exhibited by either of the materials reacting alone. Fly ash has been used extensively for stabilisation of high plasticity clays (Ferguson, 1993; Turner, 1997). Fly ash often stabilises high plasticity clays as well as lime. One of the major benefits of using fly ash for soil stabilisation in lieu of lime is bond formation between soil grains and cementitious products released due to the reaction of lime with fly ash when mixed with water (Ferguson, 1993).

### 6.3.4   Polyurethane (PU)

The use of polyurethane is a low-cost technique in soil stabilisation. The workable characteristic of polyurethane foam suits construction needs; it increases strength performance while reducing the compression index. Polyurethane (PU) is a polymer composed of organic joined by urethane links.

The method of PU foam injection is basically identical to conventional grouting, whereby the chemical resin is injected into the ground by drilling the hole, installing the packer with tube and filling up the voids between the ground soil (Figure 6.3). In addition to the use of PU, the bearing capacity of the soil is improved due to the expansion of voids between the soil particles (Sidek et al., 2015).

Polyurethane with maximum strength was used to stabilise soft clay soil under the layer of road pavement. The usage of the polyurethane as a stabiliser in the soil was conducted by

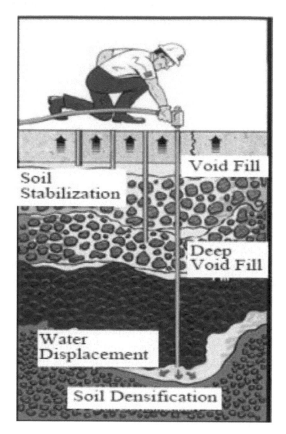

*Figure 6.3* PU injection illustration
Source: After Fakhar et al. (2016)

using the stone column method. Real field conditions of soil were used, which were with and without injection of polyurethane in order to compare the occurred settlement. The settlements that occur were monitored and observed when there were additions of different loads placed on top of the soil surface.

The parameters that were adopted for the testing in the laboratory are as follows:

• Maximum strength of polyurethane: 1:2 (4 ml)
• Moisture content: 80%
• Loading used: 10, 20, 40, 80 and 100 kg
• Recorded time: 24 hours for each loading.

The processes of soil stabilisation using polyurethane on soft clay soil were done in the lab as follows (Figure 6.4):

Phase 1: Identifying the maximum strength of the polyurethane.
Phase 2: Preparation of the clay soil on the model tank which the soil is the same condition as in the field.

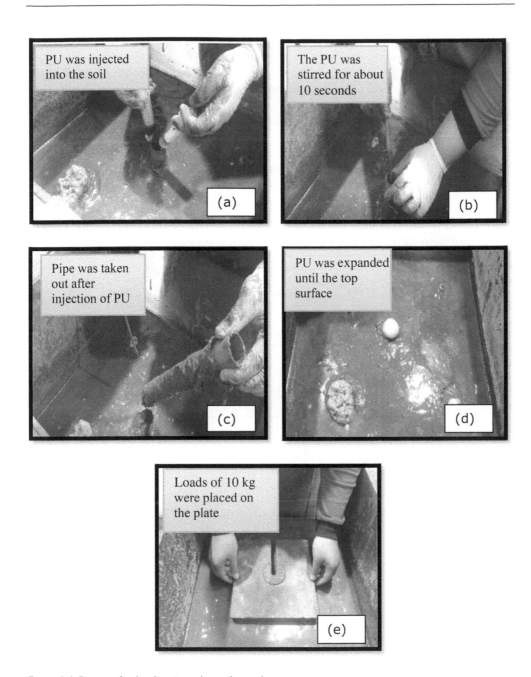

*Figure 6.4* Process for loading into the soil sample

Source: Kamaruddin (2017)

Phase 3: Samples were then left for 24 hours with the placement of wood on top of the soil surface so that the soil would settle down simultaneously.

Phase 4: Four pipes were used as a stone column for the injection of the polyurethane inside the soil. Before the pipes were taken out, the chemical had been stirred for a while so that the reaction between the chemicals would occur. Figure 6.4 shows the process of preparation for the soil sample into the box.

Phase 5: Five different loads were chosen to be placed on top of the soil surface. The load was placed on the center of the area so the distribution of the stress can occur.

Phase 6: Readings of the settlement that occurred was recorded on a 24-hour basis.

Results for the control sample and stabilised sample with polyurethane (PU) are shown in Table 6.1.

Figure 6.5 shows the results of the settlements that occurred for the control soil sample and stabilised soil sample with polyurethane for 1:2 (4 ml). Based on the results, total settlement that occurred with placement of the load for the control sample was higher (193 mm)

*Table 6.1* Results of the settlement for control and stabilised sample

| Load (kPa) | Settlement (mm) | |
| --- | --- | --- |
| | Control | Stabilised (PU) |
| 0 | 0 | 0 |
| 0.75 | 25 | 29 |
| 1.51 | 39 | 30 |
| 3.02 | 42 | 38 |
| 6.04 | 61 | 43 |
| 7.55 | 193 | 61 |

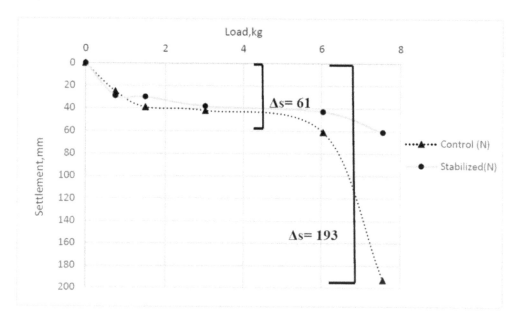

*Figure 6.5* Results of settlement for control and stabilised samples

Source: Kamaruddin (2017)

compared to stabilised (PU) (61 mm). The difference between settlements was 132 mm, which means the stabilised settlement was three times as much as the control sample. Other than that, the pattern of settlement that occurred was quite similar for the loading of 0.75–3.02 kPa, but then the pattern of the graph changed for the loading of 6.04–7.55 kPa. This may be due to the additional load that been placed on the soil.

Therefore, injection of PU with a suitable ratio on the soft clay soil was expected to reduce the settlement. Injection of PU on the soil before the road construction also will increase the strength of the soil and the road surface will perform better than a road surface without the stabiliser. Furthermore, with the implementation of the injection of polyurethane in future, it would reduce the thickness and the cost of materials that might be used on the construction of road pavement layers.

## 6.4   Modified cementitious stabilising agent

As mentioned previously, in general cement stabilisation can provide good strength. However, soil type is critical for cement stabilisation to be suitable and effective. Additives are added into cementitious chemicals to retain good strength while improving the limitations of cementitious stabilisation. Additives such as ionic and/or polymeric chemicals are used as stabilising agents.

Additives can be pre-produced and then added to mix with cementitious base chemical together with soil/aggregate materials at the site. However, this method is not advisable, as the mixing quality would be at stake. This is because it cannot be ensured that additives are mixed thoroughly with a base chemical to achieve the design composition.

Alternatively, additives can be pre-mixed, blended and batched at the factory. This method can ensure the desired chemical composition is produced during batching stage.

Typical functions of modified cementitious stabilising agent are as follows:

1.   To reduce rigidity and increase flexibility of cementitious stabilised layer and form a semi-rigid platform.
2.   To improve soaking strength over time. This function is important for application in Malaysia to resolve a low-lying area with a high water table.
3.   To improve compressibility.

Soaked CBR tests in a laboratory carried out to verify the soaking strength of crusher run stabilised using modified cementitious stabilising agent are shown at Figure 6.6. With the addition of 2% (by weight) of a renowned modified cementitious stabilising agent, 7-day soaked CBR was improved from 43.5% to 88.2%. With 3% (by weight) of the stabilising agent, the CBR value was further improved to 288%.

Apart from technical effectiveness in improving soil properties, a modified cementitious stabilising agent provides flexibility in terms of design. Since this type of stabilising agent is a mixture, the composition as well as the percentage can always be adjusted to suit requirements during the design stage.

Depending on composition, modified cementitious stabilising agents may also deliver the following functions:

1.   Faster chemical reaction for higher initial strength
2.   Lower moisture content from wet soils
3.   Water retention and shrinkage compensation against cracking
4.   Lower permeability of stabilised layer.

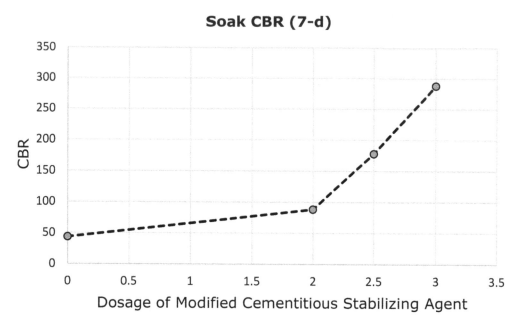

*Figure 6.6* Soaked CBR: 7 days

Source: MTS Fibromat Sdn Bhd

## 6.5   Case study

Jalan Paip is located in a residential area in Klang district, Selangor, Malaysia. Since this road serves as a short cut from Puncak Alam to Meru, traffic consists of a high volume of commercial vehicles laden up to 30 tons. Besides a high traffic volume, Jalan Paip is located in a low-lying area with a high water table. Therefore the road condition was deteriorated. Defects including differential settlement, potholes and rutting were observed along the road.

Previously, the road was repaired at various times. Methods adopted included resurfacing of asphalt concrete and mechanical reinforcement using geosynthetic products. However, defects recurred within a short period of time (Figures 6.7–6.8).

Recently, in situ chemical stabilisation was proposed to repair and rehabilitate part of Jalan Paip (Figures 6.9 to 6.12). The scope of rehabilitation included the following:

*   Remove existing asphalt concrete layer
*   Chemical stabilise and compact top 300 mm of existing road base course with renowned brand of modified cementitious stabilising agent
*   Lay asphalt concrete binder course and wearing course.

Upon completion of the chemical stabilisation process, an in situ CBR test was carried out to verify strength was achieved as per specification. Requirement was set at CBR (7-day) >120%. Although all tests were carried out before curing for 7 days, the incremental trend showed that CBR >120% was achievable (Figure 6.13).

*Figure 6.7* Existing defect: pothole
Source: Courtesy of MTS Fibromat Sdn Bhd

*Figure 6.8* Existing defect: differential settlement
Source: Courtesy of MTS Fibromat Sdn Bhd

*Figure 6.9* Existing defect: rutting
Source: Courtesy of MTS Fibromat Sdn Bhd

*Figure 6.10* Work in progress: spreading of modified cementitious stabilising agent
Source: Courtesy of MTS Fibromat Sdn Bhd

*Figure 6.11* Work in progress: in situ mixing of stabilising agent and existing road base
Source: Courtesy of MTS Fibromat Sdn Bhd

*Figure 6.12* Stabilised road base course open to traffic before laying asphalt concrete

Source: Courtesy of MTS Fibromat Sdn Bhd

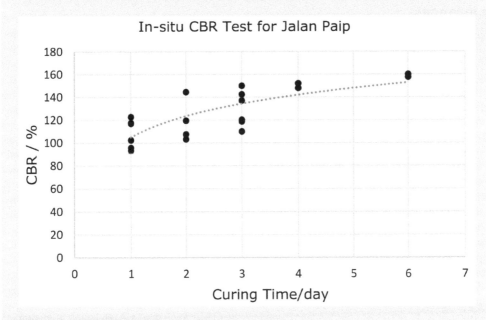

*Figure 6.13* In situ CBR test for Jalan Paip

Source: Courtesy of MTS Fibromat Sdn Bhd

The site was revisited one year after completion (Figure 6.14). No recurrence of pre-existing defects was observed. Therefore, it is concluded that using the renowned brand of modified cementitious stabilising agent in rectifying road conditions under high water table conditions with high traffic loading is effective.

*Figure 6.14* Road condition after being in use for one year

Source: Courtesy of MTS Fibromat Sdn Bhd

## References

Fakhar, A. M. M. Asmaniza, A. (2016) Road maintenance experience using polyurethane (PU) foam injection system and geocrete soil stabilization as ground rehabilitation. IOP Conference Series: Materials Science and Engineering, 136. 012004.

Ferguson, G. (1993) Use of self-cementing fly ashes as a soil stabilisation agent. Geotechnical Special Publication, No. 36. ASCE, New York, NY, pp. 1–15.

Kamaruddin, F. A. (2017) *Performance of Polyurethane on Soft Soil*. Master Thesis. Universiti Tun Hussein Onn. Malaysia.

Nikraz, H., Soomro, M. & Griffin, G. (2003) Use of limekiln dust – fly ash for stabilization of fine grained soil. In: Huat et al. (eds) *Proceedings of 2nd International Conference on Advances in Soft Soil Engineering and Technology, 2–4 July, Putrajaya*. Universiti Putra Malaysia Press, Putrajaya. pp. 41–48.

Fakhar, A.M.M. & Asmaniza, A. (2016) Road maintenance experience using polyurethane (PU) foam injection system and geocrete soil stabilization as ground rehabilitation. In: *IOP Conference Series: Materials Science and Engineering (Vol. 136, No. 1, p. 012004)*. IOP Publishing, Bristol, UK.

Sidek, N., Mohammed, K., Mohamed Jais I. B. & Abu Bakar I. A. (2015) Strength characteristics of polyurethane (PU) with modified sand. *Applied Mechanics and Materials*, 773–774, 1508–1512.

Turner, J. (1997) Evaluation of western coal fly ashes for stabilization of low-volume roads. In *Testing Soil Mixed with Waste or Recycled Materials*, ed. M. Wasemiller and K. Hoddinott. ASTM International, West Conshohocken, PA. pp. 157–171. https://doi.org/10.1520/STP15649S.

Van Impe, W. F. (1989) *Soils Improvement Techniques and Their Evolution*. A.A. Balkema, Rotterdam, The Netherlands.

# Chapter 7

# Deep stabilisation using chemical additives

## 7.1 Introduction

Soil stabilisation refers to any methods or techniques that improve the engineering properties of soil, such as shear strength, compressibility, stiffness and permeability. Hausman (1990) stated that soil deep stabilisation methods are usually done based on the following classifications: (1) mechanical modifications, (2) hydraulic modifications, (3) physical and chemical modifications, and (4) modifications by inclusions and confinement. Further, physical and chemical modifications are divided in two parts by Hausmann (1990): (a) the mixing method and (b) the injection and grouting method. The mixing method is divided in two different methods: the shallow mixing method and the deep mixing method (DMM). The injection and grouting method will be discussed in Chapter 9.

## 7.2 Deep mixing method (DMM)

The deep mixing method (DMM) is today accepted worldwide as a soil improvement method which is performed to improve the strength, deformation properties and permeability of the soil. This method, originally developed in Sweden and Japan more than 30 years ago, is becoming well established in an increasing number of countries. In Sweden and Finland, deep stabilisation techniques are quite popularly used for stabilisation of soft soil (Åhnberg et al., 1995).

The method is based on mixing binders, such as cement, lime, fly ash, chemical grouts and other additives with the soil by the use of rotating mixing tools in order to form columns of a hardening material as chemical reactions between the binder and the soil grains are developed (Costas and Chatziangelou, 2008). Based on design requirements, site conditions, soil and rock layers, restraints and economics, the use of DMM is increasingly spreading. These methods have been suggested and applied for soil and rock stabilising, slope stability, liquefaction mitigation, vibration reduction (along the railway), road and railroad and bridge foundations and embankments, construction of excavation support systems or protection of structures close to excavation sites, solidification and stabilisation of contaminated soil and so forth.

The main advantage of these methods is long-term increase in strength, especially for some of the binders used. The pozzolanic reaction can continue for months or even years after mixing: the increase in curing time results in an increase in the strength of cement-stabilised clay (Bergado, 1996).

Andromalos et al. (2000), Keller (2005), Raito (2017), SDFEC (2007) explained that DMM can be divided into (1) shallow soil mixing (SSM), which uses a single mechanical

*Table 7.1* The factors to be considered in the DMM method (after Warner, 2004)

| In the installation process | In the mixing process |
|---|---|
| The geometry of the mixing tool. | The rheological properties of the unstabilised soil and the mixture. |
| The retrieval rate. | |
| The rotation speed. | The type and amount of binder. |
| The feed pressure and the amount of air. | The in situ stress condition during the curing period. |
| The machine type and the driver. | |
| The in situ stress situation at the time of the installation. | The trials should be performed by different test methods. |

mixing auger located at the end of the drilling tool (Kelly bar); (ii) the cement deep mixing (CDM) system, which utilises a series of overlapping augers and mechanical mixing shafts; and (3) injection and grouting methods which can be considered a type of soil mixing (which will be discussed later).

Holm (1999) emphasised that DMM is the best way to improve soils and rocks, and the following characteristics of the soils and rocks are improved when these methods are employed: reduction of settlements, increase of stability, increase of bearing capacity, prevention of sliding failure, reduction of vibration, liquefaction mitigation and remediation of contaminated ground. Based on conditions such as the types of soil and rock layers, timetable of the project, location and importance of the project and the economic situation, the use of multiple-auger or single-auger deep mixing methods, jet grouting methods or a combination of several methods may be required.

The mixing process in deep mixing is complex and consists of many phases. There are several factors that influence the process and the results. There are a number of requirements relating to the test methodology and the factors to be considered, which are cited in Table 7.1 (Warner, 2004).

### 7.2.1  Shallow soil mixing (SSM)

This method has been used for improving shallow soils and seldom in deep mixing. Columns of stabilised material are formed by mixing the soil in place with a binder, and the interaction of the binder with the soft soil leads to a material which has better engineering properties than the original soil (Hebib and Farrell, 2003). The most popular binders are cement, lime and lime-cement together.

This technique causes the cement or binder to mix with soil homogeneously and thus produces higher quality soil-cement or soil-binder columns (Figure 7.1).

### 7.2.2  Cement deep mixing (CDM) system

The second method which is related to the deep mixing method (DMM) methods is the cement deep mixing (CDM) system. In this method, a series of overlapping or single augers with mechanical mixing shafts are applied. Figure 7.2 shows the Multi Auger DMM Machine (www.raito.co.jp).

CDM is normally utilised in soft soil that contains mineral soils such as clay or sand. In some conditions, where mineral soils are absent (in the soft soil), sand should be added before mixing in cement slurry. Figure 7.3a and 7.3b shows the DMM machine with single augers, and the soil stabilised with the DMM machine with one auger (www.raito.co.jp)

*Figure 7.1* The simple auger

Source: www.soilmec.com/en/technologies

*Figure 7.2* The Multi Auger DMM Machine

Source: After Raito (2017)

*Figure 7.3* (a) The DMM Machine with single augers (RAS column method), and (b) the soil stabilised with DMM Machine with one auger

Source: www.raito.co.jp

The advantages and main points of DMM method are (1) it is a drilling and mixing operation with low noise and low vibration and does not generate dust; (2) this method mixes soft soil in situ with cement slurry without any jetting; (3) because of a series of overlapping augers, it saves soil mixing time and labor while maintaining efficiency in comparison with previous methods; and (4) computer-based control and monitoring system ensures quality improvement (in some latest ones).

## 7.3    Lime columns

### 7.3.1    The chemistry of lime treatment

Lime in the form of quicklime (CaO), hydrated lime (Ca(OH)$_2$) or lime slurry can be used to treat soils. Hydrated lime is created when quicklime chemically reacts with water. When lime and water are added to a clay soil, chemical reactions begin to occur almost immediately. The hydrated lime produced by these initial reactions will subsequently react with clay particles as pozzolanic reactions. The calcium ions (Ca$^{++}$) from hydrated lime migrate to the surface of the clay particles and displace water and other ions. In this stage flocculation and agglomeration due to reduction of the plasticity index of the soil occurs, and the soil becomes friable and granular which makes it easier to work and compact (NLM, 2004).

The National Lime Association (2004) elaborated:

> when adequate quantities of lime and water are added, the pH of the soil quickly increases to above 10.5, which enables the clay particles to break down. Silica and alumina are released and react with calcium from the lime to form calcium-silicate-hydrates (CSH) and calcium-aluminate-hydrates (CAH). CSH and CAH are cementitious products similar to those formed in Portland cement. They form the matrix that contributes to the strength of lime-stabilized soil layers. As this matrix forms, the soil is transformed from a sandy, granular material to a hard, relatively impermeable layer with significant load bearing capacity. The process begins within hours and can continue for years in a properly designed system.

The stabilized soil is normally firm to hard and the texture is grainy. The behavior is essentially that of an overconsolidated clay (Balasubramaniam and Buensuceso, 1989; Balasubramaniam et al., 1990).

National Lime Association (2006) stated that soil stabilised properties with lime have been changed as follows:

1. Due to the chemical reaction between water and quicklime, soil moisture content decreases rapidly.
2. Lime has different effects on soil in a short time which may or may not be permanent and are more pronounced in soils with sizable clay content. These effects include soil plasticity reduction, increase in optimum moisture content, maximum dry density reduction, improved compactability, reduction of the soil's capacity to swell and shrink, and improved strength and stability after compaction.
3. Soil stabilised with lime has produced long-term strength and permanent reduction in shrinking, swelling, and soil plasticity with adequate durability to resist the detrimental

effects of cyclic freezing and thawing and prolonged soaking by enough amounts of clay and sizable mineralogy in the soil.

4.  The effects of lime stabilisation are typically measured after 28 days or longer but can be accelerated by increasing the soil temperature during the curing period. A soil that is lime-stabilised also experiences the effects of soil drying and modification.

The shear strength as well as other properties of the stabilised soil gradually improves with time through pozzolanic reactions, when the lime reacts with the silicates and aluminates in the clay. The pozzolanic reactions take place over many months and years. The amount of clay of the soil should therefore not be less than 20% if lime is used. The sum of the silt and clay fractions should preferably exceed 35%, which is normally the case when the plasticity index of the soil exceeds 10 (Broms, 1984). Due to increased solubility of the silicates and the aluminates in high pH (pH $\geq$ 12) and high ground temperature, the lime reaction with soil will be accelerated.

### 7.3.2    Design and calculation methods of bearing capacity of lime, lime-cement and cement columns

The lime column method has been used in other countries of the world (chiefly in Scandinavia) to provide additional bearing capacity and reduced settlements for soft clays. Lime columns are constructed in situ by intimate mixing of clay and finely pulverised unhydrated lime (CaO). As stated earlier, the mixing tool is first screwed into the soil down to a depth that corresponds to the desired length of the columns. The maximum length is at present 15 m (50 ft). The tool is then slowly withdrawn ($\sim$ 2.5 em/revolution) as unslaked lime is forced down into the soil through holes located just above the horizontal blades of the mixing tool using compressed air. Since the blades are inclined (or tilted), the stabilised soil will be compacted during the withdrawal. The resulting columns have the same diameter as the mixing tool (Figure 7.4) (Broms, 1991).

Bredenberg et al. (1999) emphasised that in the design of lime, lime-cement and cement columns, it is important to consider (a) the stability of the stabilised embankments, trenches and slopes are adequate (ultimate limit state, or ULS) as well as ultimate bearing capacity of the columns, (b) the total and deferential settlements as well as the lateral deformations are not excessive at working load (serviceability limit state, or SLS), and (c) nearby buildings as well as buried services and other structures are not damaged during installation of the columns.

### 7.3.3    Settlement prediction of lime, lime-cement and cement columns by Chai and Pongsivasathit (2010)

The most common methods to calculation settlements in lime, lime-cement and cement columns are under the assumption of the equal strain on the column and on the surrounding soil without distribution of load into the improved soil area. Baker (2000) stated that, in deep mixing improved soil, the settlement and its change in time mainly depend on the modulus of compressibility and the permeability of both the improved and unimproved soil. In Chai and Pongsivasathit's (2010) method (Figure 7.5), the consolidation settlement-time curve of the clayey subsoil is modified by a ground improvement using

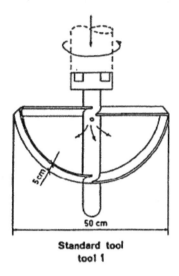

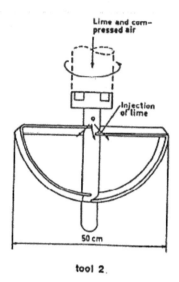

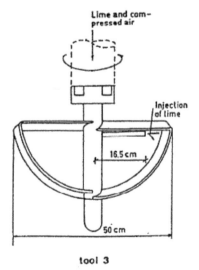

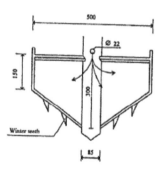

*Figure 7.4* Manufacture of lime columns, typical end mixing tools

Source: Stabilator (1998)

a floating soil-cement column. The new proposal in this method is a relative penetration of the column in the underlying soft soil, which was considered during the consolidation process.

The relative penetration of the column is influenced by the area replacement ratio ($\alpha$) and the depth replacement ratio ($\beta$), which is used to describe the relative improvement of the

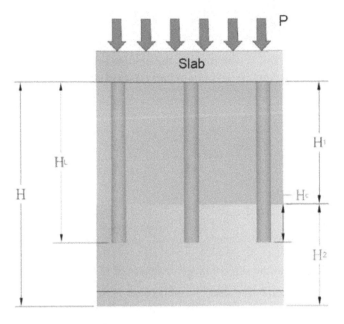

*Figure 7.5* Improved clayey subsoil by floating column

Source: After Chai and Pongsivasathit (2010)

soft soil by the column (Equations (7.1) and (7.2)). The load intensity (P) and the stiffness of the soft soil (S) also have an impact on the relative penetration (Kefyalew, 2018).

$$\alpha = \frac{d_c}{d_e} \tag{7.1}$$

$$\beta = \frac{H_l}{H} \tag{7.2}$$

$$H_c = H_l \cdot f_{(\alpha)} \cdot g_{(\beta)} \tag{7.3}$$

where

$d_c$ and $d_e$ are the diameter of the column and diameter of the unit cell which represent the column and the surrounding soil, respectively
$H$ = thickness of the soft clay soil excluding the slab thickness
$H_L$ = length of the column
$H_c$ = The thickness of the improved soil which is considered as unimproved.

Based on Chai and Pongsivasathit (2010) definition, $f_{(\alpha)}$, $g_{(\beta)}$ are bilinear functions which can be obtained as below:

$$f_{(\alpha)} = \begin{cases} \dfrac{8}{15} - \dfrac{\alpha}{75} & (10\% \leq \alpha \leq 40\%) \\ 0 & (\alpha > 40\%) \end{cases} \tag{7.4}$$

$$g_{(\beta)} = \begin{cases} 1.62 - 0.016\beta \, (20\% \leq \beta \leq 70\%) \\ 0.5 \qquad\qquad (70\% \leq \beta \leq 90\%) \end{cases} \tag{7.5}$$

The settlement of the soft soil is summing up the compression of the improved layer with thickness $H_1$ is $s_1$ and the compression of unimproved layer with a thickness of $H_2$ is $s_2$.

$$s(t) = s_1(t) + s_2(t) \tag{7.6}$$

$$(s_2) = \sum_{i=1}^{n} H_{2i} \frac{\lambda_i}{1+e_{0i}} \ln\left[1 + \frac{\Delta p_{2i}}{\sigma'_{vi}} U(t)\right] \tag{7.7}$$

$$(s_1) = \sum_{i=1}^{n} \frac{\Delta p_{1i} H_{1i} U(t)}{D_{ci}\alpha + (1-\alpha)D_{si}} \tag{7.8}$$

where

$H_{1i}$ and $H_{2i}$: thickness of the subsoil layers in layers of $H_1$ and $H_2$, respectively

$\sigma'_{vi}$: the initial vertical effective stress in sublayer O

$e_{0i}$: initial void ratio

$\lambda_i$: the slope of virgin compression line in $e - \ln(p')$ plot

$p'$: the mean effective consolidation stress.

$\Delta p_{1i}$, $\Delta p_{2i}$: the total vertical stress increments in layers $H_1$ and $H_2$, respectively.

$D_{ci}$, $D_{si}$: the constrained moduli of the column and the surrounding soil of the layer $H_{1i}$ can be calculated as:

$$D_{ci} = \frac{E_i(1-v_i)}{(1+v_i)(1-2v_i)} \tag{7.9}$$

$$D_{si} = \frac{(1-e_i)\sigma'_{avi}}{\lambda_i} \tag{7.10}$$

where $E_i$ is the elastic modulus, $v_i$ is Poisson's ratio, $e_i$ is the void ratio and is the average effective vertical stress including stress increment by the embankment of the corresponding sub-layer of the soil. For Equations (7.8) and (7.10), in using $e - \ln(p')$ it is recommended to use $k_i$ instead of $\lambda_i$ when the subsoil layer is overconsolidated. Accordingly, the final settlement (compression) can be calculated as:

$$s(t) = s_1(t) + s_2(t) \tag{7.6}$$

In this settlements prediction method, Chai and Pongsivasathit (2010) considered the improved clay subsoil as a two-layer system as shown in Figure 7.6. Then a theoretical solution proposed by Zhu and Yin (1999) was applied to estimate the degree of consolidation. In addition to the coefficient of consolidation ($C_v$), the degree of consolidation ($U$) could be influenced by the permeability ($k$) and the coefficient of volume compressibility ($m_v$) individually.

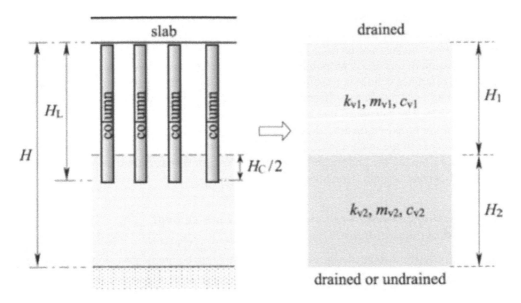

Figure 7.6 Two-layer systems for calculation of degree of consolidation

Source: After Chai and Pongsivasathit (2010)

So the value of the volume compressibility $m_{v1}$ could be evaluated using the area weighted average of the constrained moduli of the column ($D_c$) and the soil in the unit cell ($D_s$):

$$m_{v1} = \frac{1}{\alpha D_c + (1-\alpha)D_s} \tag{7.11}$$

Regarding the value of the column permeability in most cases, it is almost closer to or the same as the permeability of the surrounding soil, but due to a higher stiffness of the column, its coefficient of consolidation could be much larger than the soil in the cell. This results in a flow in the radial direction. Hence the permeability of the improved soil was determined by introducing the concept of equivalent vertical permeability of the prefabricated vertical drain. So, the value of $k_{v1}$ can be evaluated from the following equation:

$$k_{v1} = \left(1 + \frac{2.5H_1^2}{\mu d_e^2}\left(\frac{k_h}{k_v}\right)\right)k_v \tag{7.12}$$

where $k_v$ and $k_h$ are the permeability of the soft soil in the vertical and horizontal direction, respectively, and $H_1$ is the thickness of layer-1 and $\mu$ is described as follows:

$$\mu = \ln\left(\frac{n}{s}\right) + \frac{k_h}{k_s}\ln(s) - \frac{3}{4} + \frac{8H_1^2 k_h}{3d_c^2 k_c} \tag{7.13}$$

where = diameter of the smear zone and $k_c$ and $k_s$ = the coefficient of permeability of the column and the smear zone, respectively.

Hulumtaye (2017) stated that the thickness of layer-1 obtained gives a good result. But in the case of layer-2, due to large consolidation strain, its thickness is completely different before and after consolidation since it is unimproved soil. The new proposal in this case is to take the average thickness of layer-2.

The validation of this newly proposed method was done by using a finite element analysis (FEA) for a reference condition of soft clayey soil with a soil deposit 12 m thick. It is analysed using a unit cell model for different values of $\alpha$ and $\beta$ which range $10\% \leq \alpha \leq 30\%$ and $10\% \leq \beta \leq 30\%$. The calculated degree of consolidation compared with different values of $\alpha$ and $\beta$, and the proposed method shows a good prediction of the degree of consolidation. The effectiveness of the method was verified by comparing measured results from lab and case histories from the site and suggested to use for designing of a soft soil improvement by floating soil-cement column (Hulumtaye, 2017).

### 7.3.4   Bearing capacity prediction of lime, lime-cement and cement columns

If the stability is overestimated, especially for high-strength columns, then slop and bearing capacity failures have occurred. The columns could fail at lateral displacement of only 10–30 mm for 0.6 m diameter columns when the failure strain is 1%. At a failure strain of 10%, the lateral displacement at failure of the columns generally exceeds 100 mm. The permeability of the columns should be at least 300 times the permeability of the unsterilised soil for the columns to function as drains. The permeability of the stabilised soil in the columns should preferably be at least 500 to 1000 times the permeability of the unsterilised soil. Otherwise the hydraulic lag in especially long columns could affect stability (Bredenberg et al., 1999).

The bearing capacity and the shear resistance of lime, lime-cement and cement columns are governed either by the undrained or by the drained shear strength, depending on the loading rate and the permeability of the stabilised soil.

The shear strength should therefore be evaluated for both drained and undrained conditions. The lowest calculated shear resistance should be used in design. The undrained shear strength of the unstabilised soil between the columns is used to calculate the short-term stability since the shear strength increases with time due to consolidation. The lowest bearing capacity and the lowest shear strength of the columns are expected just below the dry crust, where the shear strength of the unstabilised soil usually is low as well as the total and the effective overburden pressures.

Bredenberg et al. (1999) stated that, the undrained shear strength $\tau_{fu,col}$ of soil stabilised with lime, lime-cement or cement increases with increasing confining pressure, which depends on the total confining pressure, $\sigma_h$. For an expanding cavity,

$$\sigma_h = \sigma_{ho} + c_{u,soil}\left(1 + ln\left[E_{u,soil} / 2c_{u,soil}\left(1 - v_{soil}\right)\right]\right) \qquad (7.14)$$

where $\sigma_{ho}$ is the initial total lateral earth pressure, $c_{u,soil}$ is the undrained shear strength of the unstabilised soil, $E_{soil}$ is the modulus of elasticity and $v_{soil}$ is Poisson's ratio. At

$$E_{u,soil} = 55 c_{u,soil} \text{ and } v_{soil} = 0.5 \text{ then } \sigma_h = \sigma_{vo} + 5 c_{u,soil} + m_{soil} q_o \qquad (7.15)$$

where $m_{soil}$ is the stress concentration factor for the unstabilised soil and $q_o$ is the applied unit load.

There is considerable uncertainty about the term $5c_{u,soil}$ which should only be considered after this term has been verified by field load tests. The shear strength of the unstabilised soil $c_{u,soil}$ will have about the same effect on the confining pressure and on the bearing capacity of the columns as the overburden pressure when the clay is slightly overconsolidated. The increase can be large.

The increase of the undrained shear strength caused by consolidation of the soft soil around the columns will contribute to the bearing capacity of the columns. The increases of the undrained shear strength is estimated to 5.4 kPa (0.27 × 0.2 × 5 × 20) at a $c/p'$-ratio of 0.2 and when, for example, 27% of the weight of a 5 m high embankment is transferred to the unstabilised soil between the columns. The bearing capacity of the lime columns is then increased by 81 kpa (5 × 5.4 × 3) due to the increase of the total overburden pressure at a friction angle $\phi'_{col}$ of 30 degrees when the normal pressure on the slip plane is less than about 150 kPa. When the normal critical pressure is exceeded, the increase of the bearing capacity of the columns due consolidation of the soil around the columns is estimated at 27.0 kPa $\left(\phi'_{col} = 0\right)$.

### 7.3.4.1   Short-term ultimate bearing capacity (Bredenberg et al., 1999)

When $\sigma_h \leq 100$–150 kPa, the bearing capacity can be calculated by the following equation, assuming that $\phi_{u,col} = 30°$:

$$q_{col} = q_{u,col} + 3\sigma_{vo} + 15c_{u,soil} + 3m_{soil}q_o \tag{7.16}$$

where $q_{u,col}$ is the unconfined compressive strength of the columns at the ground surface, $m_{soil}q_o$ is the increase of the overburden pressure caused by the embankment and $\sigma_{vo}$ is the initial total overburden pressure.

The bearing capacity of the columns is calculated by the following equation at $\phi_{u,col} = 0$ and $\sigma_h \geq 100$–150 kPa at the depth which corresponds to the initial total overburden pressure $\sigma_{vo}$

$$q_{col} = q_{u,col} + \sigma_{vo} + 5c_{u,clay} + m_{soil}q_o \tag{7.17}$$

The overburden pressure $(\sigma_{vo} + m_{soil}q_o)$ for the unstabilised soil next to the columns will be less than the applied unit load $q_o$ since a large part of the weight of the embankment is transferred to the columns as indicated by the stress factor $m_{soil}$.

### 7.3.4.2   Long-term ultimate bearing capacity (Bredenberg et al., 1999)

The long term ultimate bearing capacity of the columns $q'_{col}$ can be estimated from the following equation $\phi'_{col} = 30\ to\ 35°$ and $\sigma_{ho} = \sigma_{vo} + 5c_{u,col} + m_{soil}q_o$, where $\sigma_{vo}$ is the initial overburden pressure as discussed by Broms (1984). Then

$$q_{col} = 2c'_{col}\sqrt{K_P} + K_p\left(\sigma_{vo} + 5c_{u,soil} + m_{soil}q_o - U_{col}\right) + U_{col} \tag{8}$$

where $c'_{col}$ is the effective cohesion of the columns, $\sigma_{vo}$ is the initial total overburden pressure, $c_{u,soil}$ is the undrained shear strength of the unstabilised soil between the columns, $U_{col}$ is the pore water pressure in the columns and $K_p = \left(1 + sin\ \phi'_{col}\right) / \left(1 - sin\ \phi'_{col}\right)$ is coefficient of passive earth pressure.

Due to the high permeability of lime columns, it is expected that the pore water pressure in the columns for long-term conditions will correspond to the water table or to the level of the drainage layer below the embankment.

### 7.3.4.3  Residual bearing capacity

Bredenberg et al. (1999) explained that the residual bearing capacity of lime-cement and cement columns at $c'_{col,\,res} = 0$ can be calculated from:

$$q_{col,res} - U_{col} = K_p \left( \sigma_{vo} + 5c_{u,soil} + m_{soil} q_o - U_{col} \right) \tag{7.19}$$

where $K_p = \left( 1 + sin\ \phi'_{col} \right) / \left( 1 - sin\ \phi'_{col} \right)$, $\sigma'_h$ is the effective confining pressure and $U_{col}$ is the pore water pressure in the columns.

---

**Case 1**

Hamburg-Berlin Railway Line: reinforced embankment on pile-like elements

As part of upgrading the Hamburg-Berlin railway line by the German Rail Company, the Büchen Hamburg and the Paulinenaue-Friesack parts of the railway line were upgraded in 2003 to allow a train speed of 230 km/h. Because of the very soft organic soil (peat and mud) layer with insufficient bearing capacity, stabilisation of the embankment foundation was necessary at these two parts. The part of this line with a total length of 625 m was near the railway station at Büchen. The soft underground was stabilised with columns installed using the mixed-in-place (MIP) method and the embankment was reinforced with geogrids at its base on the top of the columns. The MIP belongs to the wet deep mixing methods. The underground consists of a 3–5 m fill of silty and gravely medium dense sand with slag and organic mixtures underlain by a 0.5–2 m thick layer of very soft peat and mud. The peat soil has a water content of 80%–330% and an organic content between 25%–80%. Beneath the soft layer, a medium dense and slightly silty sand layer with a thickness up to 8 m is encountered followed by boulder clay with soft to stiff consistency and a water content of 10%–20%. The MIP columns were installed using a single-axis auger. Cement slurry was continuously injected into the soil during the penetration as well as during the retrieval of the auger. Due to the rotation of the auger, the cement slurry is mixed with the soil. The MIP technique is free of vibrations and displacements and therefore had no effect on the ongoing railway traffic on the other track. The cement columns (diameter 0.63 m) were installed in a square grid of 1.5 m².

Source: Souliman and Zapata, 2011

**Case 2**

City Road Trasa Zielona in Lublin-Polen: wet deep soil mixing

The weak soil found at a depth of 3–8 m consists of loose anthropogenic fill underlain by 1–4 m thick peat and organic clay. The embankment height was 1.3–2.5 m and the equivalent live load was 30 kPa. A triangular column grid with 2 m spacing was selected, resulting in soilcrete design strength of 480–676 kPa. The required unconfined compression strength was 1.5 Mpa. Altogether, 2402 columns with a total length of 15,532 m had been constructed. The final embankment was reinforced with two layers of Tensar geogrid, resulting in the so-called load transfer platform design.

Source: Souliman and Zapata, 2011

## Case 3

Tomei Expressway expansion project

The Tomei Expressway, which connects Tokyo and Nagoya, has been an artery of Japanese culture and economy since the 17th century. Even with the addition of railroads and bullet train rails, the traffic volume along the Tomei Expressway continues to increase. Therefore, the expansion project was carried out to expand the four-lane freeway into a six-lane freeway, with three lanes in each direction. The section near Isebara, located along the foothills of the Tanzawa ridge, is underlain by consecutive sections of ridges and valleys. The ridges consist of organic clays with peat. Excessive total and differential settlement and embankment instability were expected if the new embankments were placed without improving the strength and compressibility of the organic soils. The dry jet mixing (DJM) method, which has been used in numerous projects for treating organic soils and peat with satisfactory results, was selected to treat. Staged construction at three elevations was used for soil treatment. Two mix designs were used. Lower cement injection rates, 100–120 kg cement per cubic meter of target soil, were used in the existing embankment zone to maintain the strength of soil-cement at 0.8 kgf/cm² [Editor's note: this old-metric/pre-SI unit is approximately equal to 100 kPa] – the average strength of the existing embankment soils. The higher cement injection rate, 170–230 kg cement per cubic meter of target soil, was used in the organic clays to obtain a minimum unconfined compressive strength of 7 kgf/cm² after treatment. A total of 50,215 m³ organic clay, peat and fill were treated for use as foundations of the new embankment, retaining walls and box culvert. Two sets of DJM rig were used. The deep mixing commenced in April 1994 and was completed in November 1994 without interrupting the use of the four-lane freeway. A staged construction procedure was used to perform the soil treatment within a limited working space. In addition to the foundation treatment for the new embankment, DJM also improved the foundation for the retaining wall and the box culvert and eliminated the mobilisation of pile driving equipment to the congested work zone parallel to an existing freeway.

Source: Souliman and Zapata, 2011

## References

Åhnberg, H., Johansson, S. E., Retelius, A., Ljungkrantz, C., Holmqvist, L. & Holm, G. (1995) *Cement ochKalk for DjupstabiliseringavJord, en Kemisk-FfysikaliskStudieavStabiliseringsEffekter*. Swedish Deep Stabilization Research Centre, Linköping, Swedish Deep Stabilization Research Center, Sweden, Report 48.

Andromalos, K. B., Hegazy, Y. A. & Jasperse, B. H. (2000) Stabilization of soft soils by soil mixing. *Proceedings of the Soft Ground Technology Conference*, United Engineering Foundation and ASCE Geo–Institute, Noorwijkerout, Netherlands.

Baker, S. (2000) *Deformation Behavior of LCcolumn Stabilized Clay*. Swedishdeep Stabilization Research Center, Linköping, Report 7.

Balasubramaniam, A. S. & Buensuceso, B. R. (1989) On the overconsolidated behavior of lime treated soft clay. *Proceedings of the 12th International Conference on Soil Mechanics and Foundation Engineering*. CRC Press, Rio de Janeiro, 2, 1335–1338.

Balasubramaniam, A. S., Buensuceso, B. R., Phien-Wej, N. & Bergado, D. T. (1990) Engineering behavior of lime stabilized soft Bangkok clay. *Proceedings of the 10th Southeast Asian Geotechnical Conference*. Southeast Asian Geotechnical Society, Taipei, 1, pp. 23–28.

Bergado, D. T. (1996) Soil compaction and soil stabilisation by admixtures. *Proceeding of the Seminar on Ground Improvement Application to Indonesian Soft Soils*. Jakarta, Indonesia. pp. 23–26.

Bredenberg, H., Broms, B. B. & Holm, G. (1999) *Dry Mix Methods for Deep Soil Stabilization*. CRC, Balkema, Rotterdam, The Netherlands.

Broms, B. B. (1982) Lime columns in theory and practice, Proceedings of the 12th International Conference on Soil Mechanics, Mexico, 149–165.

Broms, B. (1984) *Stabilization of Soil With Lime Columns. Design Handbook* (3rd ed.). Lime Column AB, Kungsbacka, Sweden.

Broms, B. B. (1991) Stabilization of soil with lime columns. In: Fang, H. Y. (eds) *Foundation Engineering Handbook*. Springer, Boston, MA.

Chai, J. & Pongsivasathit, S. (2010) A method for predicting consolidation settlements of floating column improved clayey subsoil. *Frontiers of Structural and Civil Engineering*, 4(2), 241–251.

Costas, A. A. & Chatziangelou, M. (2008) Compressive strength of cement stabilized soils, a new statistical model. *The Electronic Journal of Geotechnical Engineering (EJGE Journal)*, 13(B).

Hausmann, R. M. (1990) *Engineering Principles of Ground Modification*. McGraw–Hill Inc, New York.

Hebib, S. & Farrell, E. R. (2003) Some experiences on the stabilization of Irish peats. *Canadian Geotechnical Journal*, 40(1), 107–120.

Holm, G. (1999) Application of dry mix method for deep soil stabilization. In: Holm & Broms (ed) *Dry Mix Method for Deep Soil Stabilization*. CRC Press, Balkema, Rotterdam, The Netherlands.

Hulumtaye, K. (2017) *Settlement Calculation for Lime/Cement Column Improved Clay Analytical and Numerical Analyses Related to a Case Study*, Master Thesis, Royal Institute of Technology, Stockholm, Sweden.

Kefyalew, H. (2018) *Settlement Calculation for Lime/Cement Column Improved Clay Analytical and Numerical Analyses Related to a Case Study*. Royal Institute of Technology, Sweden.

Keller Company. (2005). Available from Keller: http://www.kellergrundbau.com.

National Lime Association. (NLM). (2004) "Lime Stabilization & Lime Modification," *Bulletin*, 326.

National Lime Association. (NLM). (2006) "Mixture Design and Testing Procedures Mixture Design and Testing Procedures for Lime Stabilized Soil."

Raito Company. (2017) Available from Raito: http://www.raito.co.jp/english/.

Soilmec Drilling and Foundation Equipment Company. (2007) Available from SDFEC: http://www.soil mec.it/.

Souliman, M. I. & Zapata, C. (2011) International case studies of peat stabilization by deep mixing method. *Jourdan Journal of Civil Engineering*, 5(3), 424–430.

Stabilator. (1998) *Lime Cement Columns: A New Soil Stabilization Technique*. Presented at University of Wisconsin - Milwaukee Short Course, Deep Mixing Methods, Milwaukee, WI, August 27–28, p. 129.

Warner, J. (2004) *Practical Handbook of Grouting: Soil, Rock, and Structures*. John Wiley & Sons, Hoboken, NJ.

Zhu, G. F. & Yin, J. H. (1999) Consolidation of double soil layers under depth-dependent ramp load. *Géotechnique*, 49(3), 415–421.

# Chapter 8

# Lightweight fills

## 8.1  Expanded polystyrene (EPS)

An alternative for construction on very soft soil is using the concept of the "weight credit" technique. In this technique, some thickness of the surface layer of the in situ soil is removed. If then the weight of the structure built on this site were equivalent to the weight of the soil removed, then theoretically there would be no additional stresses imposed on the in situ soil. This means as far as the underlying soil is concerned, there is no change in its state of stress, so there would be no problem with bearing capacity inadequacy, and no problem with settlement. The benefit of this concept could be further enhanced if used in conjunction with some lightweight materials (or fills). These materials are described in the following section.

It must also be noted that with the lightweight fills, because of their lightness in weight, they will exert little stress to the in situ soil. These fills could use as an option for construction on soft ground, with or without utilising the aforementioned weight credit technique.

An example of a very lightweight material is polystyrene. In block form, also known as expanded polystyrene (EPS), it can be used to cope with extreme soils, such as very soft clays and peat. A typical density of polystyrene is about 20 kg/m³, and may possibly increase to 100 kg/m³ as the material absorbs water from the ground. This is still about 20 times lighter than conventional fill such as clay and sand, and 10 times lighter than water. Figure 8.1 illustrates the principle of using polystyrene block as an alternative to heavy earth core structures built over soft ground. Because of their light weight, very little pressure is actually exerted on the existing ground, hence minimising stability and settlement problems.

The blocks are usually of dimension 2.5 m long by 1.25 m width and 0.6 m thick. The blocks are arranged until the desired height is obtained to construct, say, an embankment. The construction is then wrapped with an earth fill to act as a protector or overlaid with a concrete slab of thickness 100–150 mm.

By excavation and replacement with the polystyrene blocks, it is theoretically possible to completely float the embankment, thus imposing zero net stress to the underlying ground. This technique, as mentioned earlier, is known as "weight credit" construction. An interesting aspect of this construction is the need to have a stable water table, as any changes will alter the state of buoyancy and potentially cause movement in the system. The blocks need also to be protected from fire usually by mean of earth cover and top concrete slabs. A general review of the literature on the use of EPS is provided by Frydenlund and Aaboe (1997) and Gan and Tan (2003). Figure 8.2 illustrates some of their applications.

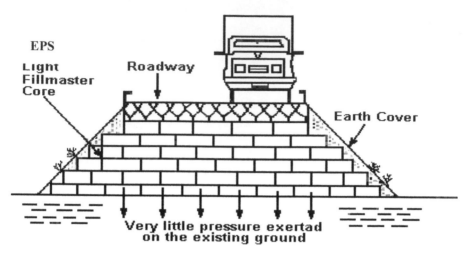

*Figure 8.1* Lightweight embankment

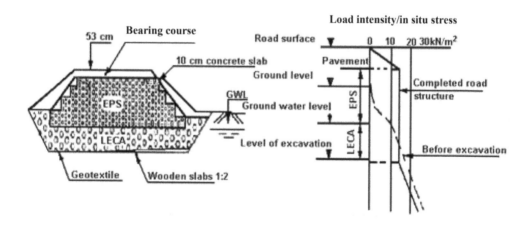

EPS – Expanded Polystyrene. LECA – Lightweight Expanded Clay Aggregates

(a)

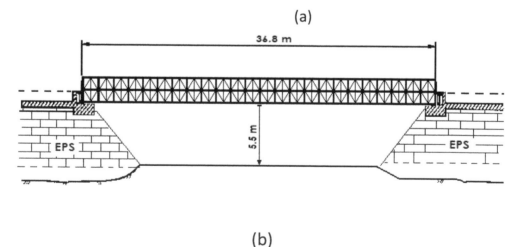

(b)

*Figure 8.2* Examples of EPS applications: (a) as compensated foundation, (b) support for bridge abutment, (c) arrangement of lightweight polystyrene blocks at culvert transition and (d) section view of culvert transition

Source: Gan and Tan (2003)

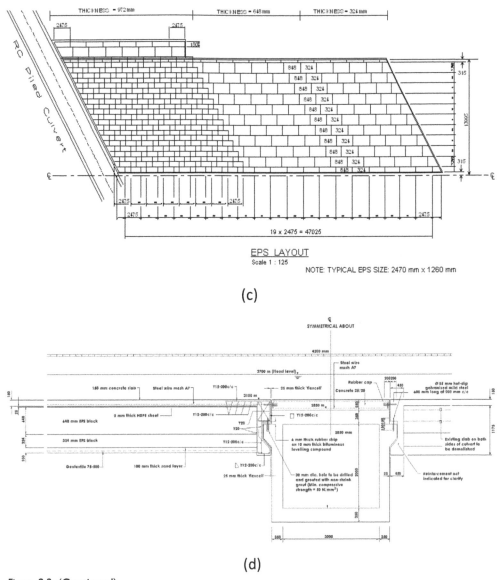

(c)

(d)

*Figure 8.2* (Continued)

Lauritzsen and Lee (2002) suggested that it is possible to use EPS as foundation for building two-story houses and gardens directly on peat in addition to roads.

## 8.2 Ultra-lightweight (bamboo culm-RPB) foundation

Very soft and often waterlogged ground, such as peat, remains a major challenge to the geotechnical community on account of its poor engineering properties. With very low bulk density (typically 0.8–1.2 Mg/m³), low undrained shear strength, very high compressibility and creep, peat deposits have very low values of allowable bearing pressure, usually not exceeding 20 kPa (Huat et al., 2014). Because of their poor engineering properties, peat deposits have tended to be avoided in earthworks projects. But when this is not possible due to a lack of adequately suitable land area for infrastructure development, a range of construction techniques can be employed, including excavation or displacement of the load-bearing peat material and its replacement with engineering fill; preloading using conventional surcharging (Beales and O'Kelly, 2008) and/or vacuum consolidation (O'Kelly, 2015); chemical stabilisation (Kazemian et al., 2009; Kalantari et al., 2011) of the peat material in situ; structural solutions (e.g., using end bearing piles with geosynthetics) that transfer the applied surface loading to underlying strata and/or the use of lightweight fill materials to reduce the magnitude of the applied loading. However, such techniques often prove uneconomical, ineffective and/or are difficult to implement in practice.

The extremely high water content and void ratio values (typically ranging between 500%–2000% and 7–30, respectively) of waterlogged peat material provide potential for buoyancy generation. Harnessing the buoyancy effect using an ultra-lightweight foundation system which uses simple technology employing green and recycled materials having relative densities lower than that of the waterlogged peat has been proposed (Ibrahim, 2017). In other words, rather than having to increase the footing area, the net bearing pressure on the soft foundation soils can be reduced overall on account of the lighter foundation construction materials employed and the buoyancy effect generated (Ibrahim, 2017). Further, by reducing the foundation settlement, the impacts on the natural groundwater regime and finely balanced eco-hydrological system of the peat bog are reduced since the reduction in the hydraulic conductivity of the underling peat is not as severe (O'Kelly, 2009). The construction materials for the proposed ultra-lightweight foundation system are bamboo culms and plastic blocks produced from compressed plastic bags (hereafter referred to as recycled plastic blocks, or RPB). The novel foundation construction incorporates the RPB within a bamboo culm frame structure that forms the foundation (footing) and which is submerged within the peat deposit (Ibrahim, 2017). Bamboo is a quick-maturing species that has been used without much difficulty as sustainable foundation construction material, especially in tropical and sub-tropical regions (Huat et al., 2014). Figure 8.3 shows the physical and numerical model of the foundation/footing system with some results as shown in Figure 8.4.

The study has shown that compared with the bamboo frame foundation, the inclusion of the RPB within the cavities of the bamboo frame greatly increases the undrained bearing capacity, with greater improvements achieved at higher water content due to increase in buoyancy of peat material and for larger volumes of bamboo frames and plastic bags on account of the increased buoyancy effect. Further, under the action of the applied load, the peat materials cannot encroach to the same extent into the bamboo frame with the RPB in place, meaning that a greater volume of bearing peat material must be displaced, leading to enhanced bearing resistance.

### 8.2.1  Design example

A stress intensity of 2500 kN (250 tonnes) is to be applied on bamboo-plastic bag frame foundation. The peat is hemic with 50% fiber content ($f$) with a bulk density of 1.030 Mg/m³ and an average moisture content of 1200%. Determine the volumes and masses of bamboo and plastic bags required for developing the foundation.

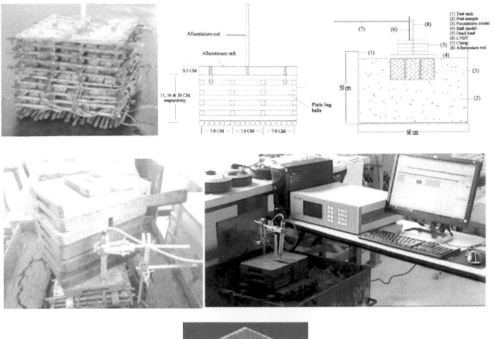

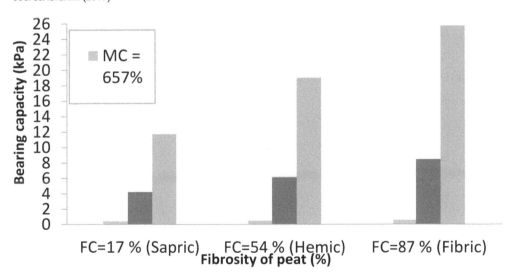

*Figure 8.3* Physical and numerical model and experimental setup of a bamboo culm – RPB foundation/ footing system

Source: Ibrahim (2017)

*Figure 8.4a* Effects of fibrosity and average water content on the undrained bearing capacity of model footing (M1)

Source: Ibrahim (2017)

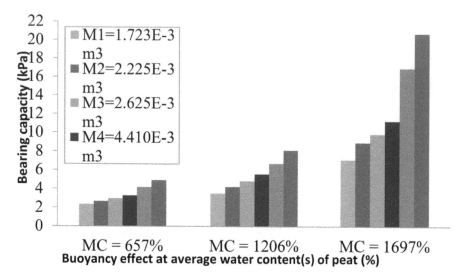

*Figure 8.4b* Effect of volumes of bamboo frame and plastic bags at lower, medium and higher water content on bearing capacity and buoyancy effect

Source: Ibrahim (2017)

### 8.2.1.1    Solution

STEP 1: GIVEN INFORMATION

$Q_a$ = 2500 kN = 250 tonnes = 2,500,000 kg m/s²
$f$ = 50%
$\rho_{peat}$ = 1.030 Mg/m³ = 1030 kg/m³

$$\rho_M = 0.466 \frac{Mg}{m^3}$$

Volume of composite foundation, $V_c$ = ?

STEP 2: DETERMINING THE VOLUME OF COMPOSITE FOUNDATION REQUIRED

From the equation:

$$Q_a = Vg(\rho_{peat} - \rho_M)$$

$V = V_c$, is given by,

$$V_c = \frac{Q_a}{g((\rho_{peat} - \rho_M)}$$

$$V_c = \frac{2,500\,(kN)}{10(m/s^2)\{(1.030 - 0.466)\,Mg/\,m^3\}}$$

$$V_c = \frac{2,500 \times 10^3\,(kg)\,(m/s^2)}{10(m/s^2)\{(1.030 - 466)\,kg/\,m^3)\}} = 443.26\,m^3$$

Therefore, the volume of composite foundation required is 443.26 m³.

STEP 3: DETERMINING THE VOLUME OF BAMBOO, $V_{BB}$ AND PLASTIC BAGS, $V_{PB}$
REQUIRED EACH

A trial and error estimation of the volume of bamboo and plastic bags is made using the equation,

$$V_{pb} = 6.2374 \ V_{bb} - 29.78$$

The values obtained are compared with the values from the design chart in Figure 8.5 below. Once the values from the equation and the chart are satisfied, and the total volume $(V_{bb} + V_{pb})$ equals the calculated volume of composite foundation, $V_c$, the estimated volumes are adopted.
    For example,

$V_{bb} = 50$, $V_{pb} = 6.2374 \ (50) - 29.78 = 282.09$
$V_{pb} + V_{bb} = 282.09 + 50 = 332.9 \ \text{m}^3 \neq 436.4 \ \text{m}^3$
At $V_{bb} = 65.36$, $V_{pb} = 6.2374 \ (65.36) - 29.78 = 377.90$
And $V_{pb} + V_{bb} = 377.90 + 65.36 = 443.26 = V_c$

$V_{bb} = 65.36$, and $V_{pb} = 377.90$ are also found to be similar to the values from the chart in Figure 8.5.
    Therefore, adopt $V_{bb} = 65.36 \ \text{m}^3$ and $V_{pb} = 377.90 \ \text{m}^3$.

STEP 4: DETERMINING THE MASSES OF BAMBOO AND PLASTIC BAGS FOR
DEVELOPING THE FOUNDATION

Mass of bamboo, $M_{bb} = r_{bb} \times V_{bb}$

$M_{bb} = 0.62 \times 65.36$
$M_{bb} = 40.52 \ \text{Mg} = 40{,}520 \ \text{kg}$

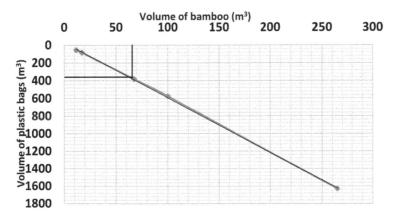

Figure 8.5 Design volume of bamboo, $V_{bb}$ against volume of plastic bags, $V_{pb}$

Mass of plastic bags, $M_{pb} = \rho_{bb} \times V_{pb}$

$M_{pb} = 0.44 \times 377.90$

$M_{pb} = 166.276$ Mg $= 166,276$ kg

The total mass of composite foundation model is $M_{bb} + M_{pb} = 206.796$ Mg $= 206,796$ kg. Therefore, the load intensity to be applied based on Archimedes' principle alone (i.e., on water) is given by;

$W_i = W_M + Q_a = 206,796 + 250,000$ kg $= 456,796$ kg $= 456.796$ tonnes

And 40,520 kg of bamboo and 166,276 kg of plastic bags are required to sustain a stress of 250 tonnes on water.

CHECK

From equation, $W_i = W_M + Q_a$

$F_{buoyant} = \rho_{peat} \times V_{peatdisplaced} \times g \geq W = W_i = W_M + Q_a$

$F_{buoyant} = 1.03 \times 443.26 \times 10 = 4,565.578$ MN $= 4,565,578$ kN $= 456,578$ kg $= 456.578$

tonnes.

$F_{buoyant} = 456.578$ kg $< 456,796$ kg $= W_i$, with a difference of 218 kg.

REMARKS

$F_{buoyant}$ is slightly less than $W_i$ due to approximations and some calculation error. However, the volume of bamboo and plastic bags needs to be increased slightly to overcome this problem.

Considering that peat has fiber materials to further enhance the capacity of the foundation, and the fiber content is 50% fiber, there is no need for increasing the volume of foundation materials at the moment.

From design chart in Figure 8.6, the applied stress supposed to have been multiplied by 1000%. Being that the ultimate stress to be applied is 250 tonnes, and there is no need for increasing the load intensity due to increase in capacity generated as a result of 50% fiber content, the volume of foundation materials is divided by 1000. Therefore, 40.52 kg of bamboo and 166.276 kg of plastic bags are required to sustain the 250-tonne stress.

Assuming a factor of safety of 2 is applied, the masses are further multiplied by 2. Thus, 81.04 kg of bamboo and 332.552 kg of plastic bags are required to sustain 250 tonnes on peat with the parameters described in the question.

STEP 5: DETERMINING THE DIMENSION OF THE FOUNDATION MODEL

For the dimension of the foundation, it could be a cuboid of equal edges (L = B = W), or otherwise.

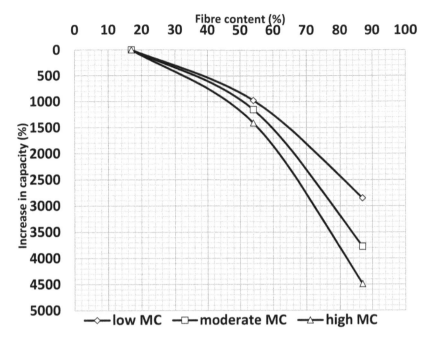

*Figure 8.6* Design chart on effects of increase in fiber contents of peat samples at low, moderate and high moisture contents

Assuming a dimension of 10 m × 10 m × 4.5 m, the volume is 450 m³ > 443.26 m³ ($V_c$). The applied stress is 25 kPa (2500/100 kN/m²).

## 8.3 Other lightweight materials

Other materials that could be considered for the "weight credit" construction technique include shredded waste tires, wood fibers or wood chips, sawdust or bales of peat.

The use of shredded waste tires or tire chips (Figure 8.7) proved to be advantageous. Its low unit weight served to limit settlement on compressible ground and the applications using this material for road projects have been successful.

Embankments constructed of wood fibers have also been found to perform well. When using wood fiber, it is recommended that the height of the wood fiber should not exceed 5 m and ventilation to the wood fiber should be reduced to prevent spontaneous combustion. Figure 8.8 shows a wood fiber–filled embankment.

In Japan, foamed cement paste (FCP), a material prepared by mixing foam with cement paste, is widely used as lightweight geomaterials for road or embankment construction (Yasuhara et al., 2003). FCP has a larger wet density (0.5–0.8 Mg/m³) compared with EPS, however it has an advantage in stability against chemicals, and its strength can be controlled by its composition.

Ghani et al. (2003) describes the potential use of waste tires shredded into chips and mix with binder in form of cement and rice husk ash (RHA) to form blocks as lightweight fill, either for embankments or backfill behind retaining walls.

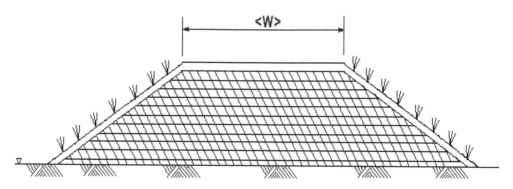

*Figure 8.7* The use of scrap tires as lightweight fill

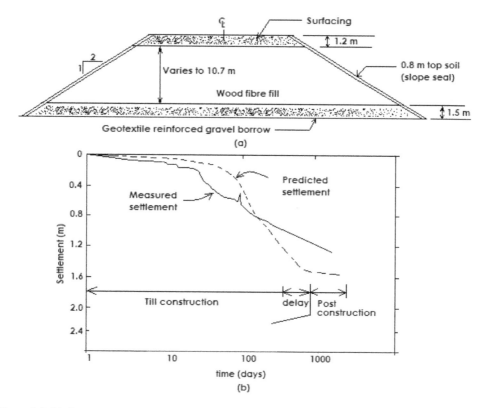

*Figure 8.8* (a) Cross-section of wood fiber–filled embankment, (b) settlement performance

## References

Beales, S. P. & O'Kelly, B. C. (2008) Performance of embankments on soft ground: A1033 Hedon Road Improvement Scheme, UK. In: Ellis E., Yu H.-S., McDowell G., Dawson A. R. & Thom N. (eds) *Proceedings of the 1st International Conference on Transportation Geotechnics*. Nottingham, UK. CRC Press/Balkema, Leiden, Netherlands, Vol. 1, pp. 369–375.

Frydenlund, T. E. & Aaboe, R. (1997). Expanded polystyrene – the light solution. In: Huat & Bahia (eds) *Proceedings Conference on Recent Advances in Soft Soil Engineering*. Samasa Press Sdn. Bhd., Kuching, Sarawak. pp. 309–324.

Gan, C. H. & Tan, S. M. (2003). Some construction experiences on soft soil using light weight materials. In: Huat et al. (eds) *Proceedings of 2nd International Conference on Advances in Soft Soil Engineering and Technology*. Universiti Putra Malaysia Press, Putrajaya, Malaysia. pp. 609–616.

Ghani, A.H.G., Ahmad, F., Hamir, R. & Mohd, S. (2003) Yielding behaviour of scrap tire base lightweight geomaterial subjected to repeated loading. In: Huat et al. (eds) *Proceedings of 2nd International Conference on Advances in Soft Soil Engineering and Technology*. Universiti Putra Malaysia Press, Putrajaya, Malaysia. pp. 681–690.

Huat, B.B.K., Prasad, A., Asadi, A. & Kazemian, S. (2014) *Geotechnics of Organic Soils and Peat*. CRC Press/Balkema, Leiden, The Netherlands.

Ibrahim, A. (2017) *Lightweight Buoyant Foundation on Peat Soil Using Bamboo Culms and Plastic Bags*. PhD Thesis (unpublished), University Putra Malaysia.

Kalantari, B., Prasad, A. & Huat, B.B.K. (2011) Stabilising peat soil with cement and silica fume. *Proceedings of the Institution of Civil Engineers – Geotechnical Engineering*, 164(1), 33–39.

Kazemian, S., Asadi, A., Huat, B.B.K., Prasad, A. & Rahim, I.B.A. (2009) Settlement problems in peat due to their high compressibility and possible solution using cement columns. In: Chen, S-E., de Leon, A. D., Dolhon, A. M., Drerup, M. J. & Parfitt, M. K. (eds) *Forensic Engineering 2009: Proceedings of the Fifth Congress on Forensic Engineering, Washington, DC*. ASCE, Reston, VA, USA. pp. 255–264.

Lauritzsen, R. & Lee, L. T. (2002). EPS – foundation for marshland Sarawak. In: Huat et al. (ed) *Proceedings of 2nd World Engineering Congress*. Universiti Putra Malaysia Press, Kuching. Sarawak. pp. 203–207.

O'Kelly, B. C. (2009) Development of a large consolidometer apparatus for testing peat and other highly organic soils. *SUO — Mires and Peat*, 60(1–2), 23–36.

O'Kelly, B. C. (2015) Case studies of vacuum consolidation ground improvement in peat deposits. In: Indraratna, B., Chu, J. & Rujikiatkamjorn, C. (eds) *Ground Improvement Case Histories: Embankments with Special Reference to Consolidation and Other Physical Methods* (1st ed.). Butterworth-Heinemann, Kidlington, Oxford, UK, Ch 11. pp. 315–345.

Yasuhara, K., Horiuchi, S. & Murakami, S. (2003). Weight reducing geo-techniques for construction of coastal structures. In: Huat et al. (eds) *Proceedings of 2nd International Conference on Advances in Soft Soil Engineering and Technology*. Universiti Putra Malaysia Press, Putrajaya, Malaysia. pp. 523–564.

# Chapter 9

# Grouting

## 9.1 Introduction

Filling the voids in the ground for increasing the shear strength and compressive strength as well as reducing the conductivity in an aquifer is done by grouting (Moseley and Kirsch, 2004). Grouting techniques are divided into hydro fracture grouting, permeation grouting, compaction grouting and jet grouting.

### 9.1.1 Hydro fracture grouting

In hydro fracture grouting, by using the grout under pressure the fractures of the soil or rock will be deliberated. Typically, it is used to compact and stiffen the ground or to access otherwise inaccessible voids, thus reducing the mass permeability of the ground and produced the controlled uplift of structures (CIRIA, 2000).

### 9.1.2 Compaction grouting

In compaction grouting, the grout is not designed to penetrate the soil voids; it displaces them. In granular soils, the volume of voids are reduced but it shouldn't be in maximum density of the soil (Karol, 2003). In compaction grouting, a very stiff (say 25 mm slump) mortar is injected into loose soil, forming grout bulbs which displace and densify the surrounding ground without penetrating the soil pores (Hausmann, 1990).

### 9.1.3 Permeation grouting

Permeation grouting is the process of filling joints in soils and rocks with grout. In this method the formation of soil and rock will not be disturbed. CIRIA (2000) stated that this technique is generally used to reduce soil permeability as well as to strengthen and stiffen the soil.

### 9.1.4 Compensation grouting

Compensation grouting is an alternating use of compaction, permeation or hydro fracture grouting. By using this method, the movement of ground will be minimised, which would affect the existing structure, particularly tunnel excavations.

### 9.1.5   Jet grouting

Jet grouting can be considered a type of soil mixing. In order to inject a liquid into voids within a structure, it is necessary to displace the gases and liquids from within them. This utilises high velocity, 28–42 MPa back pressure and jets to hydraulically shear the soil and blends a cement grout or suitable binder to form a soil-cement column or column with soil and special binder. There are four basic jet grouting systems which are widely used nowadays (Keller, 2005): (1) single phase (grout injection only), (2) dual phase (grout + air injection), (3) triple phase (water + air injection followed by grout injection) and (4) super jet grouting (air injection + drilling fluid by grout injection) (Figure 9.1).

Table 9.1 shows the methods which are commonly utilised in jet grouting systems. An explanation and comparison of each method will be made subsequently. Jet grouting systems have some similarities with the DMM methods. Apart from having the same mixing tools, this method also applies the same process whereby the in situ soil will be cut and broken by high pressure jet of slurry and produce a homogeneously improved zone

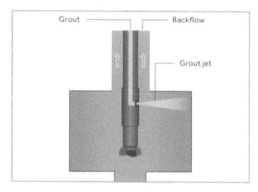

(a)  Single Direct Process

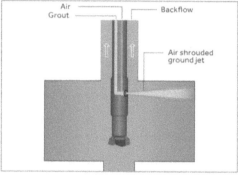

(b)  Double Direct Process

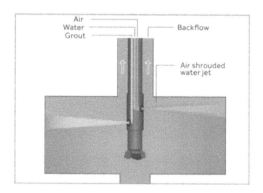

(c)  Tripple Separation Process

*Figure 9.1* Jet grouting systems for soil improvement

Source: www.keller.com

*Table 9.1* Jet grouting methods

| Method | Explanation | Example |
|--------|-------------|---------|
| Jet Grouting Systems | Single Phase | Maxperm grouting system |
| | | Navigational drilling system |
| | | Vacuum grouting injection |
| | Dual Phase | Dry jet mixing system (DJM) |
| | Triple Phase | Jumbo eco pile system (JEP) |
| | Super Jet Grouting | Ras-jet system |

around the mechanically mixed core. In addition, for underwater applications, it is desirable to have highly flowable grout that can resist water dilution and segregation and spread readily into place. The slump of concrete or grout is a good measure of the consistency and flow characteristics of a concrete or grout mixture. This equates to a mid-range slump. A very high slump grout gives maximum water dilution. A very low slump grout results in little or no flow characteristics. For underwater 3, the water-cementitious material ratio and the water reducer concentration (Khayat et al., 1996).

## 9.2 Chemical and cementation grouts

When water comes in contact with cement, three phenomena take place: (1) cement reacts with water, called hydration; (2) pozzolanic reactions between $Ca(OH)_2$ from burnt cement and pozzolanic minerals in the soil; and (3) ion exchange between calcium ions (from cement and additives) with ions present in the clay, which leads to an improvement in the strength of the treated soil (Huat et al., 2013).

Grouting and chemical grouting technologies have grown over the last few decades. Today, commercially available chemical grouts cover a wide range of materials, properties and costs and give the grouter the opportunity to select a grout for specific job requirements. Huat et al. (2013) emphasised that

> Chemical grout can be classified in single step and two step processes. In one step process, all the ingredients are premixed prior to injection, the system are designed that the reaction takes place in–situ. In the two step process, the initial chemical is injected into soil mass then follow by the second chemical material to react with the first in–situ and to stabilise the mass.

Gleize et al. (2003) stated that pozzolanic materials and fillers (kaolinite and sand) can be added to the chemical grouts as these materials change the microstructure of mortars and consequently modify some of its properties.

The difference between chemical grout and cementation grout is that the chemical grout can be used to fill the finer voids of soil particles up to 10–15 $\eta$m in diameter. In other words, it has better penetration ability than the cementitious grout (USACE, 1995). In this section, the most popular grouts will be compared together according their properties.

### 9.2.1  Sodium silicate system

Sodium silicate is the basis for many chemical grout formulations in addition to the silicate-chloride two-shot injection. It is considered free of health hazards and environmental effects (Karol, 1982), and offers specifics advantages for soil stabilisation: (1) reliable and proven performance, (2) safety and ease of use, (3) environmental acceptability and compatibility and (4) adaptability over a wide range of applications (PQ Corporation, 2003).

This grout system (i.e., sodium silicate and the reactant solution with cement) can be injected separately in two steps. The two compound system has also been used below a water table and produces a high-strength permanent grout if not allowed to dry out (Clarke, 1984; USACE, 1995; Shah and Shroff, 1999). Karol (2003) reported that when sodium silicate solution and an appropriate solution of alkali metal salts (sodium and potassium) are mixed, the reaction forming a gel is virtually instantaneous. If the silicate solution has not been moved away (by groundwater or gravitational forces), it is penetrated by thin fingers and lenses of chloride solution. As the reaction is very rapid, all the solutions cannot make contact with the soil particles.

Sodium silicate has been developed into various grout system such as silicate chloride system. Most of the systems are based on the reacting a silicate solution to form a colloid which polymerises further to form a gel that binds the soil particles. The silicate solution concentration that may be used in grouting is in the range of 10%–70% by volume, depending on the material being grouted and the desired result to achieve. For a system using amide as a reactant, the amide concentration may vary from less than 1% to greater than 20% by volume. In practice, the amide concentration ranges between 2% and 10% (USACE, 1995).

### 9.2.2  Silicate chloride amide system and silicate aluminate amide system

The silicate chloride amide system is one of the widely used silicate grout system containing sodium silicate as a gel forming material. The silicate aluminates amide system's behavior is similar to the silicate chloride amide system but is better for shutting off seepage or flow of water. The amide will act as a reactant and the calcium chloride or sodium aluminates will be used as the accelerator. The function of the accelerator is to control gel time and impart strength to the gel (Kazemian et al., 2010).

### 9.2.3  Aminoplasts

Aminoplasts consist of urea and formaldehyde. The rapid grout reaction in hot and acidic environments makes this product difficult to handle. An intermediate stage between liquid and solid urea–formaldehyde is used instead of the pure liquid phase. Aminoplasts with formaldehyde and acid catalyst contents are toxic and corrosive. In the gelled state, the aminoplast may contain leachable, unreacted formaldehyde. It is suitable for ground with a pH less than 7 (Karol, 2003; Kazemian et al., 2010).

### 9.2.4  Acrylamide

Acrylamide-based grouts come closest to satisfying the attributes of an ideal grout. They show easy penetration and maintain their initial viscosity until at the very end of the gelling stage when they rapidly set. They have good gel time control and adequate strength for most applications (Karol, 2003). The grout exhibits good penetrability, with a constant low viscosity during induction period and better gel control with adequate strength.

However, it is highly toxic and unsuitable for potable water application (Shah and Shroff, 1999). Acrylamide has a low chemical resistance toward acidity condition; therefore, it is not suitable for application in peat because peat is acidic in nature. The new acrylate gels are suitable for works that require low viscosity and a well-controlled gel time, however, the cost is higher than sodium silicates (Nonveiller, 1989).

### 9.2.5 Epoxy resins

Epoxy resins are liquid pre-polymers with a hardening agent. They usually exhibit very high tensile, compressive and bond strength. Generally epoxy resins will have either good chemical resistance or good heat resistance (Magill and Berry, 2006). The low viscosity has a better penetrability but greater shrinkage and less strength due to the weak bonding and lead to more subsidence, whereas the high viscosity may be better if adequate pressure is maintained long enough to permit the grout filling into small voids. However, the epoxy is one of the resin types which are toxic in nature, and special care is required during the handling (Rawlings et al., 2000).

### 9.2.6 N–methylolacrylamide (NMA)

N–methylolacrylamide (NMA) is inert and essentially non-toxic if properly catalysed. So, it is better than acrylamide grout. However, NMA has an extremely low viscosity of about 1–2 cP. The viscosity is similar to that of water; therefore the pumping flow rate will be same as the water. It has low stability under constant head pressure of the groundwater and is especially bad where acidic conditions and organic contaminants are present. The gel time is affected by the temperature and catalyst concentration. Acrylate grout is rarely used in geotechnical field since the gel will swell considerably in the presence of water. As a result, the strength of the grout will further reduce since existence of water will dilute the concentration of the grout (Magill and Berry, 2006)

### 9.2.7 Polyurethane

Polyurethane chemical grout is composed of two components of water-activated material called hydrophobic and hydrophilic resin. However, many other types of resin are produced based on these two resins. The viscosity of grout is very high, ranging from 300 cP to 2500 cP. The limitation is that the pH of water will affect the reactivity of grout. A higher pH value (> 7) will increase the activity of the grout. Thus it is favorable for alkaline soil and unsuitable for acidity soil, such as peat. Besides, the gel time of the polyurethane is controlled by the molecular weight, intermolecular forces and stiffness of chain units, crystallisation and cross linking (Vinson, 1970). The polyurethane is toxic in nature, so it mostly applicable in forming to block water inflow (water reactive resins).

### 9.2.8 Phenoplasts

Phenoplasts are polycondensates resulting from the reaction of a phenol on an aldehyde. There are several factors that control the phenoplast gel time including pH. For any given solution concentration, a pH slightly above 9 achieves the shortest gel time. Nonetheless, a catalyst, usually sodium hydroxide, is required to control pH. Another variable factor affecting gel time is the diluted grout concentration. Initial viscosity for field work ranges from 1.5 to 3 cP. The strength of phenoplasts is comparable to the high concentration of silicates. Phenoplasts are less sensitive to the rate of testing strain than other grouts, and their creep endurance limits constitute a greater percentage of their unconfined compression values. However, phenoplasts are toxic. The phenol, formaldehyde, and alkaline bases are all health hazards and environmental pollutants.

### 9.2.9  Lignosulfonates

Lignosulfonates are waste by-products of wood processing in paper manufacturing. Though the grout is non-toxic by itself, both in its original liquid state and dried form, the sodium dichromate additive is highly toxic (Nonveiller, 1989). If the lignosulfonate is acidic (pH < 6), no additive is required. Acids and acid salts are used only to control pH > 6 (Karol and Dekker, 1983). The grout has a viscosity range between 3 cP to 8 cP with strength comparable to acrylamide grouts (Nonveiller, 1989). However, it is highly toxic and not suitable to be used domestically.

## 9.3  Decision on choosing the grout

In order to choose a grout type, several properties of grout should be concerned, such as rheology, setting time, toxicity, strength of grout and grouted soil, stability or permanence of the grout and grouted soil, and the penetrability and water tightness of the grouted soil (Rawlings et al., 2000). Moreover, the spreading of grout plays an important role in the development of grouting technology. Magill and Berry (2006) emphasised that the comparison between chemical grouts will be made according to the penetrability of grout in soil and the range of curing time for each type of grout. In the field, the grouting method requires an extensive consideration on the grout hole equipment, distance between boreholes, length of injection passes, number of grouting phases, grouting pressure and pumping rate (Shah and Shroff, 1999). The information presented in Table 9.2 and Figure 9.2 will help in selecting the grouts for the specific requirement.

Table 9.2 Ranking based to toxicity, viscosity and strength (after Shah and Shroff, 1999, Huat et al., 2013)

| Grouts | Toxicity | Viscosity | Strength |
| --- | --- | --- | --- |
| Silicate | | | |
| Joosten process | Low | High | High |
| Siroc | Medium | Medium | Medium-High |
| Silicate-Bicarbonate | Low | Medium | Low |
| Lignosulphates | | | |
| Terra Firma | High | Medium | Low |
| Blox–All | High | Medium | Low |

| Grouts | Toxicity | Viscosity | Strength |
|---|---|---|---|
| Phenoplasts | | | |
| Terramier | Medium | Medium | Low |
| Geoseal | Medium | Medium | Low |
| Aminoplasts | | | |
| Herculox | Medium | Medium | High |
| Cyanaloc | Medium | Medium | High |
| Acrylamides | | | |
| AV–100 | High | Low | Low |
| Rocagel BT | High | Low | Low |
| Nitti– SS | High | Low | Low |
| Polyacrylamides | | | |
| Injectite 80 | Low | High | Low |
| Acrylate | | | |
| AC–400 | Low | Low | Low |
| Polyurethane | | | |
| CR–250 | High | High | High |

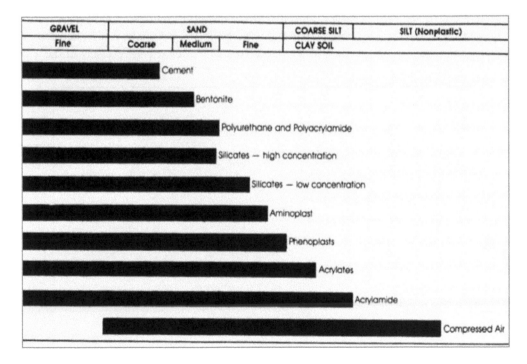

*Figure 9.2* Penetrability of various grouts

Source: Kazemian et al. (2010)

## 9.4   Case studies

### Case study 1: Soil improvement using high pressure jet grouting

Rehabilitation and upgrading of an existing motorcycle lane along the Federal Road, Petaling Jaya, Malaysia, was carried out to provide better infrastructure to the increasing numbers of road users. The length of proposed road was 16.2 km (dual carriageways) with an overhead bridge. Figure 9.3 shows the location of the site.

The overhead bridge was one of the major challenges in this case due to its proximity to a river. The length of the bridge was 120 m and the width was 5.56 m. Eight caisson piles were installed to act as foundation for the bridge column, as shown in Figure 9.4. Figure 9.5 shows engineering borelog of the site. The site comprises loose silty sand to depth of 7 m with SPT < 4, followed by a layer of medium dense to dense sand with SPT of 18–50 up to a depth of 18 m.

The main problem encountered for the construction of caisson piles up to depth 5 m was due to the collapsing of the loose sandy soil and high water table. To overcome this problem, the high-pressure jet grouting technique was used to improve soil strength during the installation of caisson piles. This is important in protecting the caisson wall from collapsing. Figure 9.6 shows a typical section of the jet grouting. Figure 9.7 shows the actual construction of the jet grouting.

The jet grouting method utilises a high-pressure hardening agent as the jetting fluid, thus ground drilling and hardener filling are conducted in one operation. The hardening agent (a water-cement mixture is normally used) is jetted horizontally together with compressed air through the rotating nozzle at the end of grout rod to fracture soil at a high pressure of 180

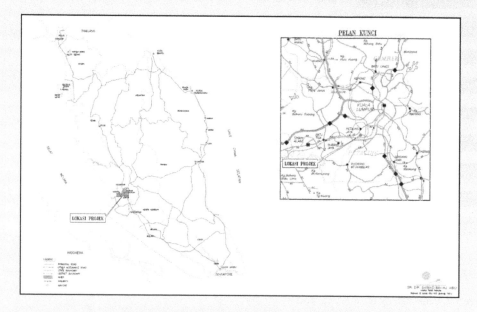

*Figure 9.3* Location of the motorcycle lane rehabilitation and upgradation project

*Figure 9.4* Location of the overhead bridge

bars to form a cylindrical column. The jet of grout is issued perpendicular to the rod, which rotates about its axis and is simultaneously withdrawn or lowered. In this way, a column of consolidated soil mass (so-called jet grout or JGP) up to 1000–1200 mm in diameter can be formed in soft soil. The following were the installation parameters adopted in this project:

For JGP 1200 mm:

- Jet grout pressure 180 bars
- Jet nozzle diameter 2.8–4.0 mm
- Withdrawal rate 16–18 min/m
- Rotation speed 5–10 rpm (rev/min)
- Air pressure flow 2–3 m³/min
- Discharge rate 0.06 m³/min
- Unconfined compressive strength ($q_u$) 10–100 kg/m²
- Modulus of elasticity (E50) 1000 or more kg/m²
- Water-cement ratio 1:1

The installation procedures adopted are as follows:

- Phase 1: Holes were drilled up to the required depth using traditional rotary drilling and water jetting method.
- Phase 2: Once the final depth had been reached, high pressure water was jetted horizontally with air jetting, which caused cutting of the ground adjacent to the jets.

- Phase 3: After cutting of the ground, grout jetting replaced water jetting. Rotation of the jet rod during withdrawal at predetermined rates resulted in an approximately circular shaped and continuous injected column (so-called JGP).

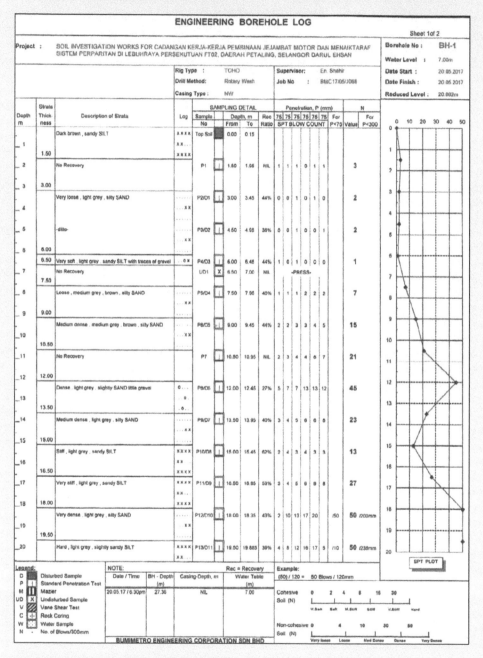

Figure 9.5 Engineering borehole log of the site

oject : SOIL INVESTIGATION WORKS FOR CADANGAN KERJA-KERJA PEMBINAAN JEJAMBAT MOTOR DAN MENAIKTARAF SISTEM PERPARITAN DI LEBUHRAYA PERSEKUTUAN FT02, DAERAH PETALING, SELANGOR DARUL EHSAN

Borehole No : BH-1
Water Level : 7.00m
Date Start : 20.05.2017
Date Finish : 20.05.2017
Reduced Level : 20.082m

Rig Type : TOHO
Drill Method: Rotary Wash
Casing Type : NW
Supervisor: En. Shahir
Job No : BMC17/05/J068

| Depth | Strata Thickness | Description of Strata | Log | Sample No | Depth, m From | Depth, m To | Rec Ratio | 75 | 75 | 75 | 75 | 75 | 75 | For P<75 | N Value | For P<300 |
|---|---|---|---|---|---|---|---|---|---|---|---|---|---|---|---|---|
| | | | XXXX | | | | | | | | | | | | | |
| 21 | 21.00 | | ..XX | | | | | | | | | | | | | |
| | | Very dense, light grey, SAND | | P14/D12 | 21.00 | 21.345 | 35% | 8 | 11 | 15 | 19 | 16 | | /45 | 50 | /195mm |
| 22 | 22.50 | | | | | | | | | | | | | | | |
| 23 | | Hard, light grey, brown, slightly silty SAND | ...XX | P15/D13 | 22.50 | 22.91 | 40% | 8 | 10 | 13 | 15 | 16 | 6 | /35 | 50 | /260mm |
| 24 | | -ditto- | | P16/D14 | 24.00 | 24.45 | 38% | 7 | 10 | 11 | 12 | 13 | 14 | /75 | 50 | /300mm |
| 25 | 25.50 | | ...XX | | | | | | | | | | | | | |
| 26 | | Very dense, brown, light grey, sandy SILT with traces of gravel | XX.X XOXX | P17/D15 | 25.50 | 25.875 | 32% | 8 | 12 | 13 | 17 | 20 | | /75 | 50 | /225mm |
| 27 | 27.00 | | XX.O | | | | | | | | | | | | | |
| | 27.36 | No Recovery | | P18 | 27.00 | 27.36 | NIL | 9 | 11 | 14 | 17 | 19 | | /60 | 50 | /210mm |
| 28 | | END OF BH-1 AT DEPTH 27.36m | | | | | | | | | | | | | | |
| 29 | | | | | | | | | | | | | | | | |
| 30 | | | | | | | | | | | | | | | | |
| 31 | | | | | | | | | | | | | | | | |
| 32 | | | | | | | | | | | | | | | | |
| 33 | | | | | | | | | | | | | | | | |
| 34 | | | | | | | | | | | | | | | | |
| 35 | | | | | | | | | | | | | | | | |
| 36 | | | | | | | | | | | | | | | | |
| 37 | | | | | | | | | | | | | | | | |
| 38 | | | | | | | | | | | | | | | | |
| 39 | | | | | | | | | | | | | | | | |
| 40 | | | | | | | | | | | | | | | | |

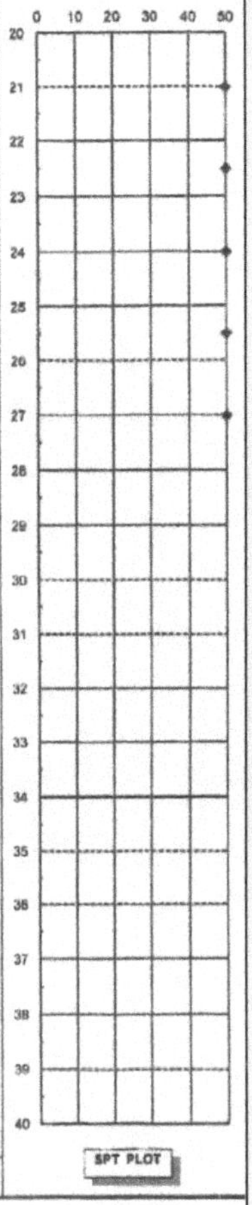


Legend:
- D Disturbed Sample
- P Standard Penetration Test
- M Mazier
- UD Undisturbed Sample
- V Vane Shear Test
- C Rock Coring
- W Water Sample
- N - No. of Blows/300mm

NOTE: Rec - Recovery

| Date / Time | BH - Depth (m) | Casing-Depth, m | Water Table (m) |
|---|---|---|---|
| 20.05.17 / 6.30pm | 27.36 | NIL | 7.00 |

BUMIMETRO ENGINEERING CORPORATION SDN BHD

Example:
(50) / 120 = 50 Blows / 120mm

Cohesive Soil (N): 0, 2, 4, 8, 15, 30 — V.Soft, Soft, M.Stiff, Stiff, V.Stiff, Hard

Non-cohesive Soil (N): 0, 4, 10, 30, 50 — Very loose, Loose, Med Dense, Dense, Very Dense

*Figure 9.5* (Continued)

During the jet grouting process, the slime was discharged through the clearance around the drilling rods by air lifting. The sludge was then channeled into a sump dug close to the work area.

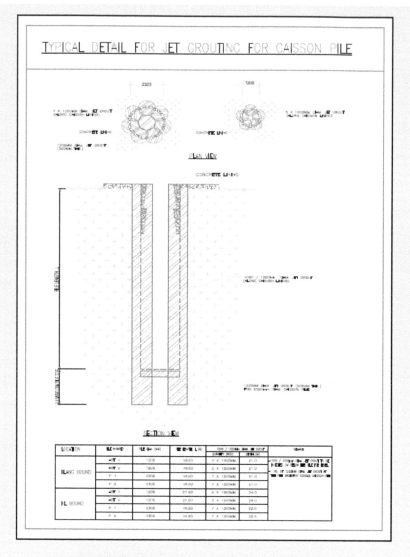

*Figure 9.6* Typical section of jet grouting

*Figure 9.7* Construction of the jet grout

Source: Photograph courtesy of HCM Engineering Sdn Bhd Malaysia, 2018

## Case study 2: Stopping gushing pipe leaks using single-component polyurethane resin

Precision Chemical Grout Company were called to a water infiltration issue in Alabama, USA. The municipality had installed PVC slip line pipes of various sizes (30-inch PVC slip line inside of 36-inch host pipe, 20-inch PVC slip line inside of 24-inch RCP host pipe, and 12-inch PVC slip line inside of 15-inch RCP host pipe, respectively). The PVC pipes were installed to stop infiltration of river and groundwater.

The monitoring reports of municipality's flow monitoring showed an active leak; in addition, they were reporting 3.5 million gallons per day (MGD) to their wastewater stream. (To put that into perspective, that is more than enough water to fill five Olympic-sized swimming pools.) The report showed using cement grout, but they were at a complete loss (Prime Resins, 2018).

By injecting the polyurethane resin in a few days to seal the annular space in seven slip line sections. This chemical grout which was used meets NSF/ANSI Standard 61 for contact with potable water. Prime Resins (2018) stated that the methods were perfected by doing some trial and error on the first section and the rest were completed with less material and in a shorter time. The method of solving the problem was by injection the polyurethane resin and using 10-foot sections of 0.5-inch PVC pipe as a probe connected to an F-valve. This allowed for standing in the invert and pushing pipe 10 feet up into the annular space, typically installing two PVC pipes for each of the leaking pipes sealed – one at 4 o'clock and one at 8 o'clock to ensure grout would fully fill the annular space between the slip line and host pipe packing open space, with towels or sandbags to contain the resulting foam. Finally, the leaks were completely sealed in six days, ending the 3.5 MGD leaks that had existed for several years with only a modest investment.

Brian A. Purcell, Gadsden (Ala.) Water Works and Sewer Board (2011), emphasised that "this particular project was an engineer's worst nightmare to develop a permanent solution to stop the leaks at the manhole connections, which resulted in substantial cost savings at our sewer treatment plant."

*Figure 9.8* From an on-site video shows the severity of the leaks

Source: After Prime Resins (2018)

# References

Brian A. Purcell, Gadsden (Ala.) Water Works and Sewer Board (2011) Available from: http://gadsd-enwater.org/

Clarke, W. (1984) Performance characteristics of microfine cement. *ASCE Geotechnical Conference*. ASCE, Atlanta, USA. pp. 14–18.

Construction Industry Research and Information Association (CIRIA), C. I. (2000) *Grouting for Grouting Engineering*. CRIRIA Press, London, UK.

Gleize, P. J., Muller, A. & Roman, H. R. (2003) Microstructural investigation of a silica fume-cement-lime mortar. *Cement & Concrete Composites*, 25, 171–175.

Hausmann, R. M. (1990) *Engineering Principles of Ground Modification*. McGraw–Hill Inc, New York.

HCM Engineering Sdn. Bhd Malaysia (2018) Available from: http://www.hcme.com.my/

Huat, B. K., Prasad, A., Asadi, A. & Kazemian, S. (2013) *Geotechnics of Organic Soils and Peat*. Taylor & Francis Group, London, UK.

Karol, R. H. (1982) Chemical grouts and their properties, In: W. H. Baker (ed), *Proceedings of the Conference on Grouting in Geotechnical Engineering*. Vol. 1, New Orleans, ASCE, New York, pp. 359–377.

Karol, R. H. (2003) *Chemical Grouting and Soil Stabilization* (3rd ed.). Marcel Dekker Inc, New York, USA.

Karol, R. H. & Dekker, M. (1983) *Chemical Grouting*. Basel Inc, New York, USA.

Kazemian, S., Huat, B. B., Prasad, A. & Barghchi, M. (2010) *A Review of Stabilization of Soft Soils by Injection of Chemical Grouting. Australian Journal of Basic and Applied Sciences*, 4(12), 5862–5868.

Keller Company. (2005) Available from Keller: http://www.kellergrundbau.com.

Khayat, K. H., Sonebi, M., Yahia, A. & Skaggs, C. B. (1996) Statistical models to predict flowability, washout resistance and strength of underwater concrete. In *Proceedings of the International RILEM Conference, Paisley, Scotland: Production Methods and Workability of Concrete* (edited by Bartos, P. J. M., Marrs, D. L., & Cleland, D. J.), E & FN Spon, 2–6 Boundary Row, London, UK.

Magill, D. & Berry, R. (2006) Comparison of chemical grout properties. *Avanti International and Rembco Geotechnical Contractors*.

Moseley, M. P. & Kirsch, K. (2004) *Ground Improvement* (2nd ed.). Taylor & Francis Group, London, UK.

Nonveiller, E. (1989) *Grouting Theory and Practice*. Elsevier Science Publication Company, New York, USA.

PQ Corporation, P. C. (2003) *Soluble Silicate in Geotechnical Grouting Applications*, Bulletin No 52–53. PQ Corporation Institute, Malvern, PA, USA.

Prime Resins, Inc. (2018) Available from: http://www.primeresins.com/

Rawlings C. G., Hellawell E. E. & Kilkenny W. M. (2000) *Grouting for Ground Engineering*. CIRIA C514, London, UK.

Shroff, A. V. & Shah D. L. (1999) *Grouting Technology in Tunneling and Dam Construction* (2nd ed.). Oxford and IBH, New Delhi and A.A. Balkema, Netherlands.

US Army Corps of Engineers (USACE), U. A. (1995) *Chemical Grouting*. US Army Corps, Washington, DC, USA.

Vinson, T. (1970) *The Application of Polyurethane Formed Plastics in Soil Grouting*. University of California, Berkeley, CA, USA.

# Chapter 10

# Other techniques

The ground improvement techniques that have been described in the first nine chapters of this book are not exhaustive. Some of the other methods such as ground freezing, soil nails and micropiles and thermal precompression are described in this chapter. In addition, new emerging technologies and ideas that might see possible application in the near future are also included here. These are alkaline activation, carbonation and application of biological methods.

## 10.1 Ground freezing method

The freezing technique is actually not a new technique. It was first used for ground improvement during the construction of the Brunkeberg tunnel in Stockholm in 1884. It is used mainly for temporary works, such as excavation in tunnel construction.

The freezing process involves constructing a frozen wall that is sufficiently impermeable on the outside face of an excavation that is to be made. Soil permeability plays no role in this process, but the natural water content of the in situ soil has to be less than about 10%.

According to Vyalov et al. (1978), shear strength and deformation parameter of the soil is significantly improved with this method. With soil with properties that do not vary much, temperature as well as efficiency to produce freezing are the most important parameters in this technique. Figure 10.1 shows the shear strength of soil due to the freezing process. The figure shows that the soil shear strength increases significantly with the freezing temperature.

There are a number of systems that can be used to produce the freezing, but the most efficient system uses liquid nitrogen. This technique is shown in Figure 10.2. The minimum temperature that can be produced is −196°C.

The width of the area that will be frozen around the freezing pipes can be estimated by taking into account the in situ groundwater flow, type of soil, thermal properties of soil with reducing temperature and the distance between the freezing pipes.

A high freezing speed can prevent the soil from expanding during freezing. However, this expansion process cannot be avoided altogether. In must be noted that water will expand by about 9% when frozen. The shear strength of clayey soil will reduced during the thawing process because excess pore water pressure will be generated in this process (Nixon and Mongestern, 1973). The freezing method is also not suitable in cases where groundwater flow is rather high, where it will give a negative effect to the freezing process. In this technique, ground around the freezing zone has to be provided with a separator zone or curtain, better still with piles or sheet piles. Figure 10.3 shows an example of an application of this technique.

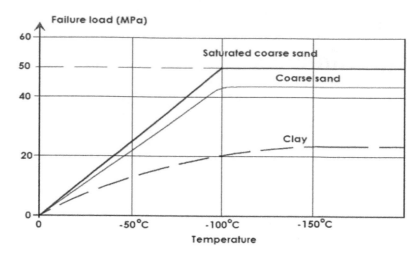

Figure 10.1 Shear strength of soil due to the freezing process
Source: Vyalov et al. (1978)

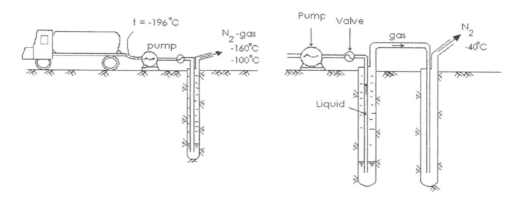

Figure 10.2 Freezing system

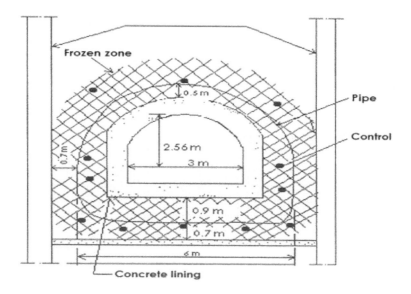

Figure 10.3 Example of an application freezing method

## 10.2   Soil nail and micropiles

Soil nail or soil nailing is an in situ reinforcement technique which consists of inserting long rods (or "nails") into undisturbed natural soil to stabilise the soil mass. Soil nails normally are installed with a near horizontal orientation (i.e., inclination of 10–20° below horizontal) and are primarily subjected to tensile stresses. Soil nailing differs from the tied-back system (ground anchor) in that soil nailing is a passive system while the tied-back system is a post-tensioned or active system, as illustrated in Figure 10.4.

In Malaysia for example, soil nails are typically consist of 16–32 mm diameter high-yield steel reinforcement bars of 6–24 m long at spacing of about 1–2 m with design load of 50–200 kN. The nails are typically inclined slightly downwards (10–20°) and grouted in pre-drilled holes having diameters in the range of 100–150 mm.

Since soil nailing is a passive system, it requires some soil deformation to mobilise resistance. Soil nails may not be appropriate for applications where very strict deformation control is required. Conceptually, soil nails are very similar to geotextile walls or reinforced earth walls (Schlosser, 1991). The main difference is in the construction process. Soil nails are constructed from top to bottom (top down), whereas geotextile or reinforced earth walls are constructed upwards.

The origins of soil nails can be traced to a support system for underground excavations in rock referred to as the New Austrian Tunneling Method (NATM). This tunneling method consists of the installation of passive steel reinforcement in the rock (e.g., rock bolts) followed by the application of reinforced shotcrete. This concept of combining passive steel reinforcement and shotcrete has also been applied to the stabilisation of rock slopes since the early 1970s. One of the first applications of soil nails was in 1972 for a railroad widening project near Versailles, France, where an 18 m high cut slope in sand was stabilised using soil nails (Ortigao and Palmeira, 2004). Because the method was cost-effective and the construction faster than other conventional support methods, an increase in the use of soil nails took place in France and other areas in Europe.

Soil nails are particularly well suited for ground conditions that require vertical or near-vertical cuts where top-to-bottom construction method is advantageous compared to other retaining wall systems. For certain conditions, soil nails offer a viable alternative from the viewpoint of technical feasibility, construction costs and construction duration when

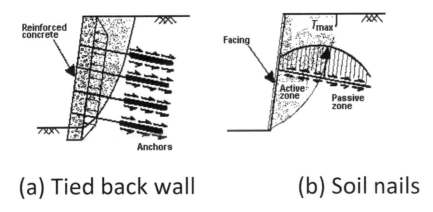

## (a) Tied back wall        (b) Soil nails

*Figure 10.4* Comparison soil nail and tied-back wall

*Figure 10.5* Soil nails for slope repairs in Malaysia

Source: http://gnpgroup.com.my

compared to ground anchor walls. They have been used successfully in highway cuts, for the repair, stabilisation and reconstruction of existing retaining structures and to stabilise natural slopes.

Alternatively, soil nails are used to stabilise landslides. In this case, the reinforcement is installed almost vertically and perpendicular to the base of the slide. In this application, the nails are installed in a closely spaced pattern approximately perpendicular to the nearly horizontal sliding surface, and subjected predominantly to shear forces arising from the landslide movement. An illustration of this application is shown pictorially in Figure 10.5.

The Geotechnical Engineering Office (GEO) of Hong Kong extensively uses soil nails to stabilise man-made slopes in residual and saprolitic soils. The only types of soils in which soil nailing cannot be applied are very loose sands and very soft clays. Most failures in residual soils take place in shallow depths, and soil nails together with deep and surface drainage can be a very economical solution. However, there are a few cases in which the failure surface is so deep that it may be more appropriate to employ long soil anchors.

Soil nails exhibit numerous advantages when compared to a rigid wall system that involves ground anchors and alternative top-down construction techniques. Some of these advantages are described below:

1.  Construction

    *   Requires smaller ROW (right of way) than ground anchors, as soil nails are typically shorter.
    *   Less disruptive to traffic and causes less environmental impact compared to other construction techniques.

- Provides a less congested bottom of excavation, particularly when compared to braced excavations.
- There is no need to embed any structural element below the bottom of excavation as with soldier beams used in ground anchor walls.
- Installation of soil nails is relatively rapid and uses typically less construction materials than ground anchor walls. Shotcrete is also a rapid technique for placement of the facing.
- Easy adjustments of nail inclination and location can be made when obstructions (e.g., cobbles or boulders, piles or underground utilities) are encountered.
- Overhead construction requirements are smaller than those for ground anchor walls because soil nails do not require the installation of soldier beams; this is particularly important when construction occurs under a bridge.
- Soil nails are advantageous at sites with remote access because smaller equipment is generally needed.

2. Performance

- Soil nails are relatively flexible and can accommodate relatively large total and differential settlements.
- Measured total deflections of soil nails are usually within tolerable limits.
- Soil nails have performed well during seismic events owing to overall system flexibility.

3. Cost

- Soil nails are more economical than conventional concrete gravity walls when conventional soil nail construction procedures are used.
- Soil nails are typically equivalent in cost or more cost-effective than ground anchor walls when conventional soil nailing construction procedures are used. Cost of soil nails can be only 50% of a tied-back wall.
- Shotcrete facing is typically less costly than the structural facing required for other wall systems.

4. Design reliability

- Saprolitic soils frequently present relict weak surfaces which can be undetected during site investigation. Soil nailing across these surfaces can lead to an increased factor of safety and increased reliability, as compared with other stabilisation solutions.

There is, however, one major limitation of the soil nailing technique, which is excessive wall deformation. Soil nailing mobilises soil strength and the soil mass deforms, leading to displacements in the surroundings of the wall. This can bring unacceptable deformation to a sensitive structure in the vicinity of the wall. Placement of the shotcrete requires that the excavated faces be freestanding for a period of time. Corrosion protection therefore requires careful attention in aggressive environments.

*Design considerations*: As recommended in Geotechnical Engineering Circular No.7 by the US Federal Highway Administration, the analysis and design of soil nails must consider two distinct limiting conditions: strength limit states and the service limit states.

*Strength limit states*: These limit states refer to failure or collapse modes in which the applied loads induce stresses that are greater than the strength of the whole system or individual components, and the structure becomes unstable. Strength limit states arise when one or more potential failure modes are realised. The design of a soil nail should ensure that the system is safe against all of the potential failure conditions as follows:

1. External failure mode
2. Internal failure mode
3. Facing failure mode.

Table 10.1 summarises the minimum recommended factor of safety for the design of soil nail using the allowable stress design method.

*Table 10.1*  Minimum Recommended Factors of Safety for the Design of Soil Nail Walls Using the ASD Method[1] (FHWA, 2015)

| Limit State | Condition | Symbol | Minimum Recomm. Factors of Safety, Static Loads | Minimum Recomm. Factors of Safety, Seismic Loads |
|---|---|---|---|---|
| Overall | Overall Stability | $FS_{OS}$ | 1.5[2] | 1.1[6] |
| Overall | Short Term Condition, Excavation | $FS_{OS}$ | 1.25-1.33[3] | NA |
| Overall | Basal Heave F | $FS_{BH}$ | 2.0[4], 2.5[5] | 2.3[5] |
| Strength-Geotechnical | Pullout Resistance | $FS_{PO}$ | 2.0 | 1.5 |
| Strength-Geotechnical | Lateral Sliding | $FS_{LS}$ | 1.5 | 1.1 |
| Strength-Structural | Tendon Tensile Strength (Grades 60 and 75) | $FS_T$ | 1.8 | 1.35 |
| Strength-Structural | Tendon Tensile Strength (Grades 95 and 150) | $FS_T$ | 2.0 | 1.50 |
| Strength-Structural | Facing Flexural | $FS_{FF}$ | 1.5 | 1.1 |
| Strength-Structural | Facing Punching Shear | $FS_{FP}$ | 1.5 | 1.1 |
| Strength-Structural | Headed Stud Tensile (A307 Bolt) | $FS_{FH}$ | 2.0 | 1.5 |
| Strength-Structural | Headed Stud Tensile (A325 Bolt) | $FS_{FH}$ | 1.7 | 1.3 |

Notes

1 The limit state and symbol nomenclature differ from that presented in the previous version of this manual. Many of these changes reflect the move toward using LRFD terminology as presented in AASHTO (2014).

2 For non-critical, permanent structures, some Owners may accept a design for static loads and long-term conditions with FSOS = 1.35 when uncertainty is considered to be limited due to the availability of sufficient geotechnical information and successful local experience on soil nailing.

3 This range of safety factors for global stability corresponds to the case of temporary excavation lifts that are unsupported for up to 2 days before nails are installed. The larger value may be applied to critical structures or when more uncertainty exists regarding soil conditions.

4 This factor of safety for basal heave is applicable to permanent walls for short-term conditions.

5 This factor of safety for basal heave is applicable to permanent walls for long-term conditions.

6 The minimum FSOS for seismic overall stability should be 1.0, when horizontal seismic coefficients are used and these were derived from estimated, allowable seismic deformations.

*Service limit states*: These limit states refers to conditions that do not involve collapse but rather impair the normal and safe operation of the structure. The major service limit state associated with soil nails is excessive wall deformation. Other service limit states include total or differential settlements, cracking of concrete facing, aesthetics and fatigue caused by repetitive loading.

There are other important factors that, if not properly addressed during design, can result in problems during operation. Two of these additional factors are drainage of the soil behind the system and corrosion of the soil nail bar or other metallic components. Corrosion is a long-term effect that has to be considered in relation to strength limit states, as corrosion affects the tensile capacity of soil nails. Corrosion of soil nail bars can lead to excessive deformations and, in an extreme case, can cause the eventual collapse of the system.

*External failure modes*: External failure modes refer to the development of potential failure surfaces passing through or behind the soil nails (i.e., failure surfaces that may or may not intersect the nails). Similar to reinforced structures, during analysis of external failure modes, the soil nail mass is generally treated as a block. Stability calculations take into account the resisting soil forces acting along the failure surfaces to establish the equilibrium of this block. If the failure surface intersects one or more soil nails, the intersected nails contribute to the stability of the block by providing an external stabilising force that must be added to the soil resisting forces along the failure surface. The evaluation of external stability is an important aspect in the design of soil nail because the magnitude and consequence of failure can be significant. External stability analyses are performed to verify that the proposed soil nail is able to resist the destabilising forces induced by the excavation and service loads for each of the potential failure modes.

Factors that control external stability include system height; soil stratigraphy behind and under the system; width of the nailed zone (i.e., soil nail lengths); and soil, nail and interface strengths. The following external failure modes are considered in the analysis of soil nail systems (Figure 10.6).

1. Global failure mode
2. Sliding failure mode (shear at the base)
3. Bearing failure mode (basal heave).

The global stability of soil nails is commonly evaluated using two-dimensional limit-equilibrium principles, which are used for conventional slope stability analyses. In limit-equilibrium analysis, the potentially sliding mass is modeled as a rigid block, global force and/or moment equilibrium is established, and a stability factor of safety that relates the stabilising and destabilising effects is calculated. As with traditional slope stability analyses, various potential failure surfaces (i.e., planar, wedge, log spiral, circular) are evaluated until the most critical surface (i.e., the one corresponding to the lowest factor of safety) is obtained. Global stability analyses are performed using computer programs either for conventional slope stability analysis or specifically developed for the design of soil nails.

Table 10.2 summarises some of the available methods for analyzing soil nail structure available in the literature. All methods divide the soil mass behind the face into an active and a passive zone (Figure 10.7), separated by a slip surface. The global stability analysis considers the stabilising effect of nails acting on the slip surface. However, they differ with respect to the shape of the slip surface, the forces assumed to act on a nail, and the method of calculation of stability.

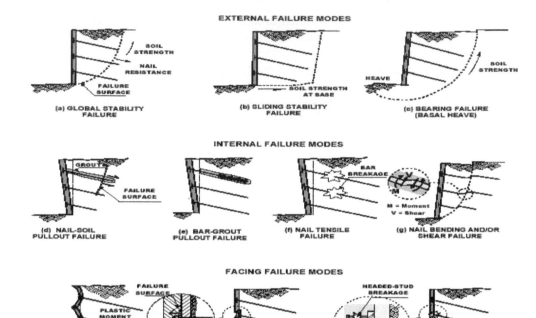

Figure 10.6 Modes of failure of soil nails

Source: FHWA (2003)

Table 10.2 Methods of analysis of soil nailed structures

| Characteristics | Methods | | | | | |
|---|---|---|---|---|---|---|
| | German | Davis | Multicriteria | Kinematical | Cardiff | Yield |
| Analysis | Limit equilibrium | Limit equilibrium | Limit equilibrium | Internal stresses | Limit equilibrium | Yield theory |
| Division of soil mass | Two wedges | Two blocks | Slices | | Slices | Rigid block |
| Factor of safety | Global | Global | Global and local | Local | Global | Global |
| Failure surface | Bi-linear | Parabolic | Circular or polygonal | Log spiral | Log spiral | Log spiral |
| Nails resist to: | | | | | | |
| Tension | X | X | X | X | X | X |
| Shear | | | X | X | X | |
| Bending | | | X | X | X | |
| Wall geometry | Vertical or inclined | Vertical | Any | Vertical or inclined | Vertical or inclined | Vertical or inclined |
| No of soil layers | I | I | Any | I | I | I |

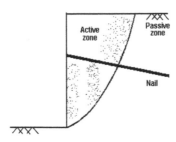

*Figure 10.7* Active and passive zones

The main shortcoming of limit equilibrium-based methods is that they do not provide a prediction of deformations, nor do they consider the deformation required to mobilise the resisting forces in the soil and soil nails.

Sliding stability analysis considers the ability of the soil nail to resist sliding along the base of the retained system in response to lateral earth pressures behind the soil nails. Sliding failure may occur when additional lateral earth pressures, mobilised by the excavation, exceed the sliding resistance along the base. Concepts similar to those used to assess sliding stability of gravity retaining structures (in which the Rankine or Coulomb theories of lateral earth pressures are used) can be applied to assess the sliding stability of a soil nail system. Again, the soil nail system is modeled as a rigid block against which lateral earth forces are applied behind the retained soil. The rigid block here is defined by a nearly horizontal surface through the base of the wall (or slightly below the base, if a weak, horizontal seam of soil is present), extends behind the nails, and exits with a steep angle at the ground surface behind the reinforced zone. It is assumed that the displacements of the soil block along its base are large enough to mobilise the active pressure behind the block. The factor of safety against sliding is calculated as the ratio of horizontal resisting forces to the applied driving horizontal forces.

Bearing capacity, though not very often, may be a concern when a soil nail system is excavated in fine-grained, soft soils. Because the system facing does not extend below the bottom of the excavation, the unbalanced load caused by the excavation may cause the bottom the excavation to heave and trigger a bearing capacity failure of the foundation.

*Internal failure modes*: Internal failure modes refer to failure in the load transfer mechanisms between the soil, the nail, and the grout. Soil nails mobilise bond strength between the grout and the surrounding soil as the soil nail system deforms during excavation. The bond strength is mobilised progressively along the entire soil nail with a certain distribution that is affected by numerous factors. As the bond strength is mobilised, tensile forces in the nails are developed. Depending on the soil nail tensile strength and length, and the bond strength, bond stress distributions vary and different internal failure modes can be realised. Typical internal failure modes (Figure 10.6 above) related to the soil nail are:

1. Nail pullout failure: A failure along the soil-grout interface due to insufficient intrinsic bond strength and/or insufficient nail length.
2. Slippage of the bar-grout interface: The strength against slippage along the grout and steel bar interface is derived mainly from mechanical interlocking of grout between

the corrugated nail bar surface. Mechanical interlocking provides significant resistance when threaded bars are used and is negligible in smooth bars. The most common and recommended practice is the use of threaded bars, which reduces the potential for slippage between the nail bar and grout.

3.  Tensile failure of the nail: The nail can fail in tension if there is inadequate tensile strength. A tensile failure of a soil nail takes place when the longitudinal force along the soil nail is greater than the nail bar tensile capacity. The nail bar capacity is function of the nail bar cross-sectional area and nail bar yield strength. The tensile capacity provided by the grout is disregarded due to the difference in stiffness (i.e., modulus of elasticity) between the grout and the nail.

4.  Bending and shear of the nails: Soil nails work predominantly in tension, but they also mobilise stresses due to shear and bending at the intersection of the slip surface with the soil nail (Schlosser, 1991). The shear and bending resistances of the soil nails are mobilised only after relatively large displacements have taken place along the slip surface. Due to this relatively modest contribution, the shear and bending strengths of the soil nails are conservatively disregarded in most of the design.

*Facing connection failure modes*: The most common potential failure modes at the facing-nail head connection are as follows (Figure 10.6).

1.  Flexure failure: This is a failure mode due to excessive bending beyond the facing's flexural capacity. This failure mode should be considered separately for both temporary and permanent facings.

2.  Punching shear failure: This failure mode occurs in the facing around the nails and should be evaluated for both temporary and permanent facings.

3.  Headed-stud tensile failure: This is a failure of the headed studs in tension. This failure mode is only a concern for permanent facings.

*Preliminary design:* Preliminary design phases may employ stability charts to evaluate required nail density. Charts developed by the French research program (Ortigao and Palmeira, 2004) may be used for its simplicity. These charts are limited to vertical wall system and relate nail density $d$ with the stability number $N$ and the friction angle ø.

Nail density $d$ is given by

$$d = \frac{\pi D q_s}{\gamma S_v S_h} \tag{10.1}$$

where $D$ is nail diameter, $q_s$ is the unit friction, $\gamma$ is soil unit weight and $s_h$ and $s_v$ are horizontal and vertical nail spacing, respectively.

The charts are used in the following way:

1.  Select chart as a function of the ratio L/H (nail length/wall height), between 0.6 and 1.2.

2.  Obtain the stability number $N = \dfrac{c}{\gamma H}$, where $c$ is soil cohesion.

3.  Determine the location of point $M$ in the selected chart with co-ordinates (tan ø, $N$).

4.  Select in the chart the appropriate value of $d$ or interpolate to obtain the required *FS*.

5.  The *FS* value is given by the ratio $FS = \dfrac{OM}{OA}$, where OM and OA are line segment lengths.

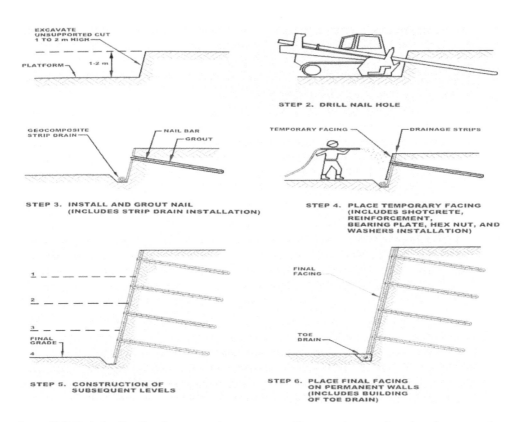

EXCAVATE
UNSUPPORTED CUT
1 TO 2 m HIGH

PLATFORM          1·2 m

STEP 2.  DRILL NAIL HOLE

GEOCOMPOSITE
STRIP DRAIN          NAIL BAR
                    GROUT

TEMPORARY FACING          DRAINAGE STRIPS

STEP 3.  INSTALL AND GROUT NAIL
        (INCLUDES STRIP DRAIN INSTALLATION)

STEP 4.  PLACE TEMPORARY FACING
        (INCLUDES SHOTCRETE,
        REINFORCEMENT,
        BEARING PLATE, HEX NUT, AND
        WASHERS INSTALLATION)

FINAL
FACING

FINAL
GRADE

TOE
DRAIN

STEP 5.  CONSTRUCTION OF
        SUBSEQUENT LEVELS

STEP 6.  PLACE FINAL FACING
        ON PERMANENT WALLS
        (INCLUDES BUILDING
        OF TOE DRAIN)

*Figure 10.8* Typical soil nail wall construction sequence with top-to-bottom (top-down) construction method

The construction method is described below and shown schematically in Figure 10.8.

Step 1: Excavation: Initial excavation is carried out to a depth for which the face of the excavation has the ability to remain unsupported for a short period of time, typically on the order of 24–48 hours. The depth of the excavation lift is usually 1–2 m and reaches slightly below the elevation where nails will be installed. The width of the excavated platform or bench must be sufficient to provide access to the installation equipment.

Step 2: Drilling nail holes: Holes are drilled to a specified length, diameter, inclination, and horizontal spacing from this excavated platform. Short nails, usually up to 3 m in length, are driven by percussion employing hand operated pneumatic hammers. This method however is inadequate in soils containing boulders, in most hard residual soils and in all permanent structures, because the steel bar is unprotected against corrosion. Another disadvantage is the resulting low soil-nail friction. The most common technique is similar to the installation of soil anchors, that is, by pressure grouting. This process leads to reasonably high unit soil-nail friction in which values greater than 100 kPa can easily be obtained in most soils. A range of equipment types can be used for installing nails, from medium sized pneumatic drill rig used for drilling a borehole in residual clays to light drill rig which can be disassembled into small pieces and easily carried out to the working platform on the top of a slope.

Step 3: Nail installation and grouting: Nail bars are placed in the pre-drilled hole. The bars are most commonly high tensile steel bar and some are galvanised to prevent erosion. HDPE or PVC centralisers are placed around the nails at typically 1.5 m c/c prior to insertion to help maintain alignment within the hole and allow sufficient protective grout coverage over the nail bar. A grout pipe (tremie) is also inserted in the drill hole at this time. When corrosion protection requirements are high, corrugated plastic sheathing can also be used to provide an additional level of corrosion protection. The drill hole is then filled with cement grout through the tremie pipe. The grout is commonly placed under gravity or low pressure. If hollow self-drilling bars are used (only as temporary structures), the drilling and grouting take place in one operation.

Step 4: Construction of temporary shotcrete facing - Prior to construction of facing element, geocomposite drainage strips are installed on the excavation face approximately midway between each set of adjacent nails. The drainage strips are then unrolled to the next wall lift. The drainage strips extend to the bottom of the excavation where collected water is conveyed via a toe drain away from the soil nail system. A temporary facing is then constructed to support the open-cut soil section before the next lift of soil is excavated. The most typical temporary facing consists of a lightly reinforced shotcrete layer commonly 100 mm thick. The reinforcement typically consists of wire mesh, which is placed at approximately the middle of the facing thickness (Figure 10.9). The length of the wire mesh must be such that it allows at least one full mesh cell to overlap with subsequent wire mesh panels. Following appropriate curing time for the temporary facing, a steel bearing plate is placed over the nail head protruding from the drill hole. The bar is then lightly pressed into the first layer of fresh shotcrete. A hex nut and washers are subsequently installed to secure the nail head against the bearing plate. The hex nut is tightened to a required minimum torque after the temporary facing has sufficiently cured. This usually requires a minimum of 24 hours. If required, testing of the installed nails to measure deflections (for comparison to a pre-specified criterion) and proof load capacities may be performed prior to proceeding with the next excavation lift. Before proceeding with subsequent excavation lifts, the shotcrete must have cured for at least 72 hours or have attained at least the specified 3-day compressive strength. Shotcrete can be applied through dry or wet mix. For small jobs, the dry mix is preferred.

Step 5: Construction of subsequent levels: Steps 1 through 4 are repeated for the remaining excavation lifts. At each excavation lift, the vertical drainage strip is unrolled downward to the subsequent lift. A new panel of wire mesh is then placed overlapping at least one full mesh cell. The temporary shotcrete is continued with a cold joint with the previous shotcrete lift. At the bottom of the excavation, the drainage strip is tied to a collecting toe drain.

Step 6: Construction of a permanent facing: After the bottom of the excavation is reached and nails are installed and load tested, a final facing may be constructed. Final facing may consist of cast-in-place reinforced concrete, reinforced shotcrete, or prefabricated panels. The reinforcement of permanent facing is conventional steel bars or wire mesh. When cast-in-place concrete and shotcrete are used for the permanent facing, horizontal joints between excavation lifts are avoided to the maximum extent possible.

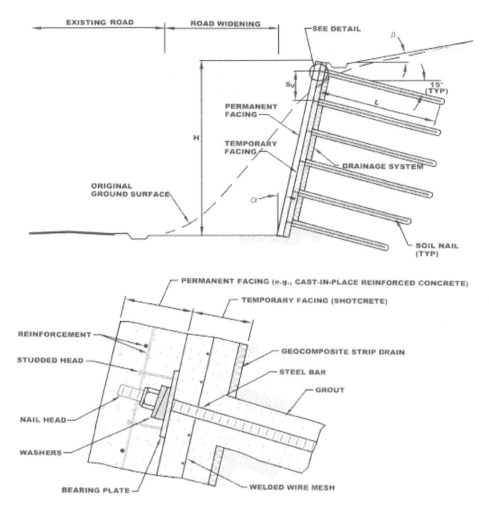

*Figure 10.9* Typical cross-section of a soil nail wall showing details of the nail head

In recent years, there has been a considerable progress in the use of steel fiber reinforced shot-crete (SFRS). Fibers are high tensile strength steel elements 30–50 mm in length and 0.5 mm in diameter with hooked ends that are mixed in the concrete as an aggregate with a dosage in the range of 35–60 kg/m$^3$. It can be used in either dry or wet sprayed concrete mix. Fibers have no effect in the compressive strength of the concrete, but increase ductility, enabling to take into account flexure tensile strength. The final SFRS product is a homogeneous material with increased crack and corrosion resistance. SFRS saves labor for mesh placement and saves total concrete volume in relation to mesh-reinforced shotcrete. SFRS complies with soil or rock surface irregularities, saving total concrete volume, as compared to the use of a steel mesh.

Vegetation can also be combined with soil reinforcement by means of nails to provide an environmentally friendly solution.

Figure 10.10 shows an example of application of the micropiles, which is also known as root piles. The micropile is initially used according to the patent of an Italian firm, Fondedile.

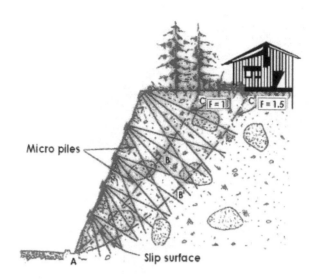

*Figure 10.10* Micropile method

Piles are inserted inside holes drilled using high-speed rotary drilling method. During the drilling, steel casings are used to prevent collapse of the drill hole as well as to limit loosening of the soil. After the drilling is completed, steel rods (usually with twisted surface), is inserted into the drill hole. The hole is then filled with cement mortar while the casing is gradually withdrawn from the earth.

However, the withdrawal of the casing sometimes caused problems. Because of this, in some practices, such as company Soletanche, the casings are left inside the soil, thus forming part of the reinforcement material.

To estimate length of required the micropile, Lizzi (1978) proposes the following simple approach. With reference to Figure 10.10, two slip surfaces, ABC, and AB'C', of the unstabilised slope are determined. The factors of safety are F = 1.0 and F = 1.5, respectively. Micropiles have to be installed at least beyond the surface of AB'C ', that is potential slip surface with F = 1.5. This determines the minimum length of the pile. Average shear strength along ABC ($R_g$) and shear strength of the micropile, $R_p$ has to be at least 1.5 times larger than active force along ABC, $P_a$. That is:

$$\left(R_g + R_p\right) / P_a > 1.5 \tag{10.2}$$

Lizzi also propose the following equation to calculate $R_p$:

$$R_p = n\, D^2 c_u \qquad (MN) \tag{10.3}$$

where

  $n$ = number of micropile
  $D$ = diameter of micropile (m)
  $c_u$ = undrained shear strength of soil (MN/m²).

## 10.3  Thermal precompression

One novel method to improve a highly organic soil such as peat is the thermal compression technique. According to Fox and Edil (1994), peat compresses more rapidly under load when heated. Two trial embankments, one (a) unheated, the other (b) heated by circulating water with temperature of 65°C flowing through heating wells installed in the peat soil, as shown in Figure 10.11. Heating causes a drastic increase in strain rate and a corresponding rapid decrease in the peat void ratio. Once the soil is cooled, it has a substantially reduced void ratio and creeps very slowly. In addition the soil develops a significant quasi-preconsolidation effect and has a very low compressibility with respect to further loading.

## 10.4  Alkaline-activated binders

Lime, cement, or lime-cement (i.e., a mixture of lime and cement) can be considered as traditional binders for soil stabilisation. Incorporation of these cementitious binders has gained popularity due to their robustness, easy adaptability and cost-effectiveness (Prusinski and Bhattacharja, 1999; Miura et al., 2001). Two main reactions which produce cementitious materials occur during soil lime-cement treatment: the hydration reaction and the pozzolanic reaction (Bell, 1996; Lorenzo and Bergado, 2006). In cemented soil, when the pore water of the soil makes contact with cement, hydration of the cement occurs rapidly and the major hydration (primary cementitious compounds) produces calcium silicate hydrate (CSH), calcium aluminate hydrate (CAH), and hydrated lime $Ca(OH)_2$ (Janz and Johansson, 2002). In lime-stabilised soil, soil particles become closer and the soil is treated through flocculation and pozzolanic reactions (Kamon and Nontananandh, 1991; Bell, 1996). Although the type of reaction in cemented soil is completely different in comparison with lime-stabilised soil, the final products, based on Si and Ca compounds, are very much alike. In terms of mechanical strength, cement-based binders usually deliver substantially better results than lime-based binders (Janz and Johansson, 2002). However, there is a concern about these

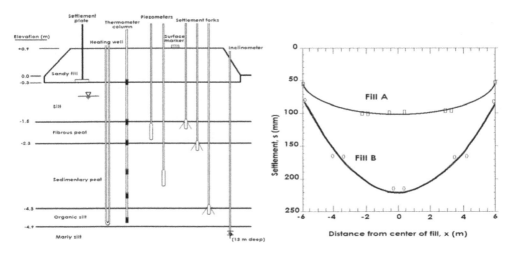

*Figure 10.11* Cross-section and performance of heated and control fill

Source: Fox and Edil (1994)

calcium-based due to their negative environmental effects during manufacture ($CO_2$ emissions) and also their cost. In addition, these binders when used in deep mixing projects show poor tensile and flexural strength and a brittle behavior (Sukontasukkul and Jamsawang, 2012; Correia et al., 2015; Pourakbar and Huat, 2016).

Alkali-activated binders, on the other hand, can constitute an interesting option to fully eliminate the use of traditional cementitious since calcium is not essential in any part of an alkali-activated structure. Alkaline activation has a history starting from the 1940s which was first demonstrated by Purdon (1940), and its application as a binder in the construction industry started in Ukraine in the 1960s (Glukhovsky, 1965). The theoretical basis of the alkaline activation system was established for the first time in 1979 by the French researcher Davidovits (1979), who introduced the term "geopolymer" to designate a new class of three-dimensional cross linked chain and pioneered the use of calcium-free systems based on metakaolin and waste materials. The synthesis of alkali-activated binders which are formed by the reaction of any Si-Al prime material, involves the dissolution of mineral aluminosilicates, hydrolysis of Al and Si components, and condensation of specific Al and Si component. Through alkali activation, when the pH falls below 13.5 using a strong alkaline solute, due to the dissolution of the source binder, the polycondensation process occurs quickly. Then groups of reactions occur in the final product of undissolved aluminosilicate component in a three-dimensional amorphous aluminosilicate. With the development of reactions, mix water is gradually consumed and a well-structured aluminum silicate hydrate (ASH) framework is formed (Weng and Sagoe-Crentsil, 2007; Yunsheng et al., 2008). The process of stabilisation in the alkaline activation and calcium-based binders is vastly different, as they use totally different reaction pathways in order to attain structural integrity. Utilisation of calcium-based binders depends on the presence of CSH and CAH gels for matrix formation and strength (Basha et al., 2005; Harichane et al., 2011). Whereas in alkaline activation, along with high alkaline solutes, a rich Si-Al resource is needed to start the dissolution and subsequent polycondensation process (Duxson et al., 2007).

A number of studies have been conducted to investigate the environmental benefits of alkaline-activated binder, notably that of Weil et al. (2009) who assessed the $CO_2$ emissions (kg $CO_2$eq/m³) and financial cost of alkali-activated binders. Those researchers reported that for an alkali-activated binder comprising 15.9% source binder dosage in the presence of high-alkali solutes, a reduction of 131 kg $CO_2$eq/m³ was achieved. Moreover, a large and growing body of literature has investigated the mechanism of the alkaline activation from wide variety of aluminosilicate source materials (Davidovits, 1988; Khale and Chaudhary, 2007; Pourakbar et al., 2015, 2016; Fasihnikoutalab, Asadi et al., 2017; Fasihnikoutalab, Pourakbar et al., 2017).

The reaction mechanism outlines the key processes occurring in the transformation of a solid aluminosilicate source into a synthetic alkali aluminosilicate can be seen in Duxson et al. (2007). Though presented linearly, these processes are largely coupled and occur concurrently. Dissolution of the solid aluminosilicate source by alkaline hydrolysis (consuming water) produces aluminate and silicate components. It is important to note that the dissolution of solid particles at the surface resulting in the liberation of aluminate and silicate (most likely in monomeric form) into solution has always been assumed to be the mechanism responsible for conversion of the solid particles during geopolymerisation. Once in solution the component released by dissolution are incorporated into the aqueous phase, which may already contain silicate present in the activating solution. A complex mixture of silicate, aluminate and aluminosilicate component is thereby formed. Dissolution of amorphous aluminosilicates is rapid at high pH, and this quickly creates a supersaturated aluminosilicate solution. In concentrated solutions this results in the formation of a gel, as the oligomers in the aqueous phase form large networks by condensation. This process releases the water that was nominally consumed during dissolution. As such, water plays the role of a reaction medium, but resides within pores in the gel. This type of gel structure is commonly referred to as bi-phasic, with the aluminosilicate binder and water forming the two phases. The time for the supersaturated aluminosilicate solution to form a continuous gel varies considerably with raw material processing conditions and solution composition and synthesis conditions. Despite this, some systems never gel. These are typically dilute, and the concentration of dissolved silicon and aluminum is observed to oscillate due to the slow response of the system far from equilibrium. After gelation the system continues to rearrange and reorganise, as the connectivity of the gel network increases, resulting in the three-dimensional aluminosilicate network commonly attributed to geopolymers.

The following factors influence the alkaline activation process in soil stabilisation:

- Curing time
- Curing temperature
- Water content
- Kaolinite content
- Alkaline concentration
- Precursor content
- Silicate/aluminate ratio
- pH
- Type of activators.

In the alkaline activation, two sources are needed for the synthesis of alkali-activated binder, namely the source materials, also known as prime material or feedstock, and the alkaline activators, which can be liquid or solid and need water to dissolve. The most common alkaline activators are a mixture of potassium or sodium hydroxide (KOH, NaOH) with sodium silicate ($nSiO_2Na_2O$) or potassium silicate ($nSiO_2K_2O$). Other alkaline minerals such as calcite, dolomite and sodium orthophosphate have also been used as parts of alkaline activators (Bakharev et al., 1999; Yip et al., 2008).

In the alkaline activation process, reactions occur at a high rate when the alkali activator contains soluble silicate such as sodium silicate (Nematollahi et al., 2014). However, the sodium silicate is not a suitable activator solution due to the environmental impact resulting from its manufacture. In fact, the production of sodium silicate involves the calcination of quartz sand and sodium carbonate ($Na_2CO_3$) at a temperature of around 1500°C, producing large amounts of $CO_2$ as a secondary product (Fawer et al., 1999; Habert et al., 2011; McLellan et al., 2011). Additionally, given that the sodium silicate used was in liquid form, it would not be applicable to dry soil mixing (Sargent et al., 2013). In such conditions, only the use of alkaline hydroxides in the alkaline activation process can be considered as a solution (Shaikh, 2013).

In the alkaline activation process, the characteristics of the source material including the particle size distribution, contents of glassy phase, reactive silicon (Si), aluminum (Al) and the presence of iron (Fe), calcium (Ca) and inert particles are of utmost importance (Khale and Chaudhary, 2007; Shi et al., 2011). The source materials are essentially agro or industrial waste materials which make this technique potentially attractive in promoting sustainable and environmentally friendly materials. They are:

- *Fly ash*, which is a fine powder which is a by-product from burning pulverised coal in electric generation power plants.
- *Palm oil fuel ash (POFA)* is by-product obtained by burning of fibers, shells and empty fruit bunches as fuel in palm oil mill boilers.
- *Rice husk ash* is a by-product taken from rice mill process, with approximately the ratio of 200 kg per ton of rice, even in high temperature it reduces to 40 kg.
- *Metakaolin* is produced by heating kaolin to a temperature of 1200–1650°F.
- Ground granulated blast furnace slag (GGBS) is obtained by quenching molten iron slag from a blast furnace in water or steam, to produce a glassy, granular product that is then dried and ground into a fine powder.

Recent research has shown that alkaline-activated stabilised soil with 10%–25% of palm oil fuel ash (POFA) and 10M KOH, and NaOH activator showed a significant increase in unconfined compressive strength of the soil as shown in Figure 10.12.

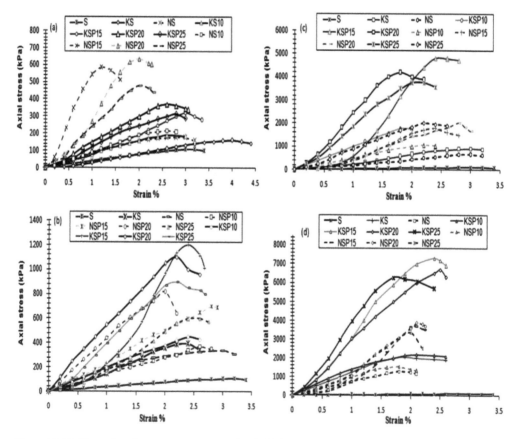

Figure 10.12 Unconfined compressive strength of alkaline-activated stabilised soil with palm oil fuel ash (POFA) with KOH and NaOH through 7, 28, 90 and 180 days of curing time

Source: Pourakbar (2015)

*Figure 10.13* SEM images of the alkaline-activated stabilised soil
Source: Pourakbar (2015)

According to this study, curing condition, type and quantity of the alkaline activators were also shown to have significant strengthening effects on the treated soil.

The SEM images (Figure 10.13) show that gaps between soil particles are filled with alkaline-activated POFA through a well-formed Si-O-Si three-dimensional structure.

Alkaline-activated stabilised soil still demonstrates a brittle behavior. To improve the sta-bilised soils brittleness, fibers can be added (Pourakbar et al., 2016; Pourakbar and Huat, 2016). The following fibers can be used.

**Artificial fibers**

- Glass fiber
- Wollastonite fiber
- Basalt fiber
- Polypropylene fiber

**Natural fibers**

- Coir fiber
- Cotton fiber
- Sisal fiber
- Raffia fiber

Figure 10.14 shows images of a wollastonite fiber alkaline-activated stabilised soil. The wollastonite fibers act similarly to plant roots in distributing the stresses over a broader area and inhibiting fissure propagation, thereby resulting in higher strength and ductility of the treated soils. In this respect, the inclusion of fiber reinforcement in alkali-activated mixtures, regardless of their particular activator-precursor combinations, increased the peak stress and improved the post peak behavior, namely by modifying the original brittle response into a more ductile one.

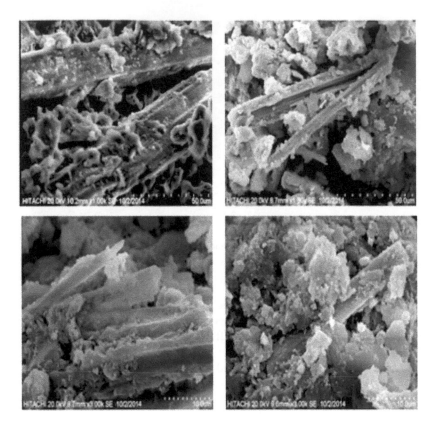

*Figure 10.14* Alkaline-activated stabilised soil with wollastonite fiber

Source: Pourakbar (2015)

As mentioned earlier, gypseous soil is a problematic soil especially in the arid and semi-arid region of the world. The soil is very sensitive toward water. When the gypsum in soil is dissolved and leached, a huge change in volume will occur causing uneven settlement. Recent study done at Universiti Putra Malaysia showed that it possible to stabilise (improve) gypseous soil using the alkaline-activated binder.

Cement and lime are the most common stabilisation materials by chemical means for gypseous soils. Cement and lime, however, as mentioned before have several shortcomings. Firstly from an environmental perspective (e.g., carbon dioxide emissions). Secondly, the availability of gypsum in gypseous soil causes sulfate attack in cement because of the presence of $Ca(OH)_2$ resulting in low durability. Sulfate attack causes expansion in cement and reduces the compressive strength with cracks and that leads to increase in soil collapsibility index.

The study showed that the alkaline activation method with fly ash class F as precursor and KOH as activator could stabilise the soils effectively (Alsafi, 2017). The collapse index decreased, and the mechanical performance of the treated soils is markedly improved. The microstructural analyses confirmed the durability of the stabilised mass. The study is important as it confirmed that the alkaline activation method has a good potential for stabilising gypseous soil.

The unconfined compressive strength of a gypseous soil with 45% gypsum treated with alkaline-activated binder (30% fly ash class F, with 8–12M KOH) showed a significant increase in the soil unconfined compressive strength as shown in Figure 10.15.

The soil collapse potential and permeability were significantly reduced as shown in Figures 10.16 and 10.17, respectively.

The SEM image (Figure 10.18) captured the formation of the ASH gel binder with a dense texture and strong matrix in G45 samples stabilised with 12M KOH-activated fly ash.

The resultant Si/Al ratios were higher than 2 in samples activated with KOH, which can be related to full geopolymerisation of fly ash and production of the ASH chains, with several peaks of quartz and no ettringite, as shown in Figure 10.19.

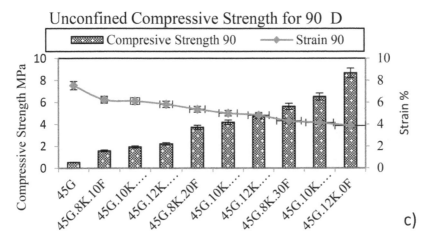

Figure 10.15 Compressive strength of treated gypseous soils samples at 90 days

Source: Alsafi (2017)

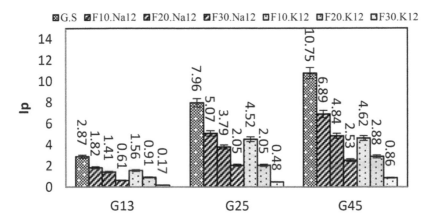

Figure 10.16 Collapsibility index of treated gypseous soil sample of 13%–45% gypsum with 10%–30% fly ash and 12M KOH and NaOH at 90 days

Source: Alsafi (2017)

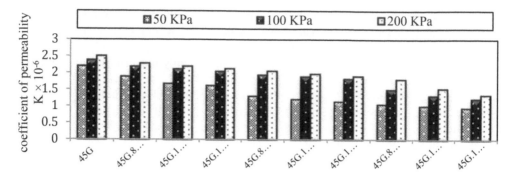

Figure 10.17 Coefficient of permeability of treated gypseous soil sample of 45% gypsum with 10%–30% fly ash and 8–12M KOH at 90 days

Source: Alsafi (2017)

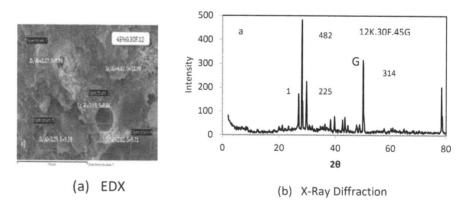

Figure 10.18 SEM of gypseous soil sample of 45% gypsum treated with 30% fly ash and 12M KOH

Source: Alsafi (2017)

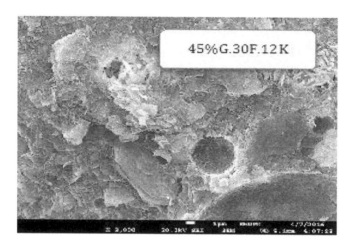

(a)  EDX                    (b)  X-Ray Diffraction

Figure 10.19 Microstructures of gypseous soil sample of 45% gypsum with 30% fly ash and 12M KOH

Source: Alsafi (2017)

## 10.5   Carbonation

The potential of stabilising soil with materials which not only help to strengthened the soil but with the ability observe/sequestrate $CO_2$ from the atmosphere is very interesting indeed, at least from the aspects of the environment. Olivine is an example of such material. Olivine occurs in both mafic and ultramafic igneous rocks and as a primary mineral in certain metamorphic rocks. Magnesium-rich olivine crystallises from magma that is rich in magnesium and low in silica. Magma crystallises to mafic rocks such as gabbro and basalt. Ultramafic rocks such as peridotite and dunite can be residues left after extraction of magmas, and typically they are more enriched in olivine after extraction of partial melts. Olivine is usually green in color and has composition that typically range $Mg^{2+}SiO_4$ and $Fe^{2+}{}_2SiO_4$. Mineral olivine is thus a type of nesosilicate or orthosilicate. It is a common mineral in the Earth's subsurface but weathers quickly on the surface. Figure 10.20 below shows the physical structure of olivine and how Mg and Fe connected to Si. The following chemical formula shows the reaction of olivine with water and $CO_2$.

$$\left(Mg, Fe\right)_2 SiO_{4(S)} + 4H_2O_{(l)} + 4CO_{2(g)} \rightarrow 2\left(Mg^{2+}, Fe^{2+}\right)_{(aq)} + 4HCO_{3^-(aq)} + H_4SiO_{4(aq)}$$

Recent research has shown that adding olivine to soil can significantly enhance the soil compressive strength at different curing time, as shown in Figure 10.21.

The SEM microstructures of the stabilised soil verified that the soil matrix was relatively dense with no greater discontinuity than that of untreated soil (Figure 10.22a). XRD analyses identified of some elements such as $Mg(OH)_2$ or brucite, MSH gel, serpentine and quartz as a result of hydration of olivine and interaction with soil as shown in Figure 10.22b.

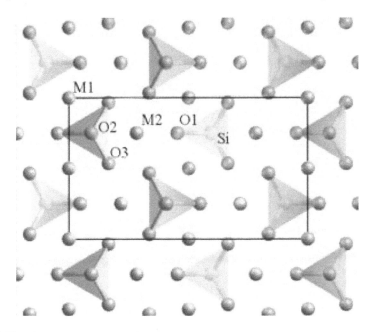

*Figure 10.20* The olivine constrictions of crystal

Source: https://en.wikipedia.org/wiki/Olivine

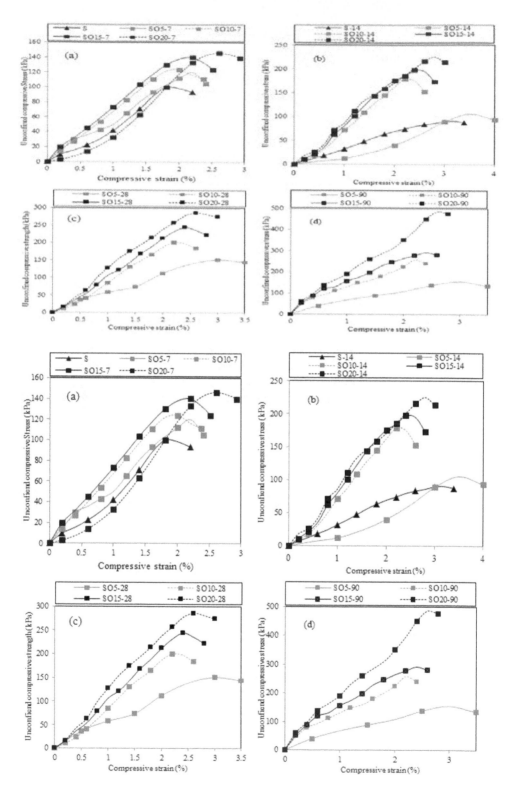

*Figure 10.21* Unconfined compressive strength of soil with 5%, 10%, 15% and 20% olivine through 7, 14, 28 and 90 days of curing time

Source: Fasihnikoutalab (2016)

(a)

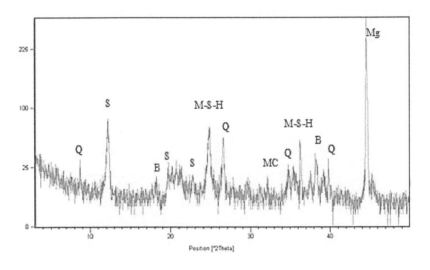

(b)

*Figure 10.22* (a) The SEM image of 20% olivine treated soil after 90 days of curing time, (b) XRD analysis of 20% olivine after 90 days of curing time

Source: Fasihnikoutalab (2016)

The following formula shows how olivine reacts with water and breaks the chemical bonding of Mg and SiO.

$$XMgO.ySiO2.zH2O \ (s) \rightarrow xMgO \ (s) + y \ SiO2 \ (s) + zH2O$$

The compressive strength of the treated soil can be further enhanced through injection of $CO_2$ under pressure, as shown in Figure 10.23. Higher strength is obtained with higher applied $CO_2$ pressure and longer carbonation period. Figure 10.24 shows the microstructure analyses.

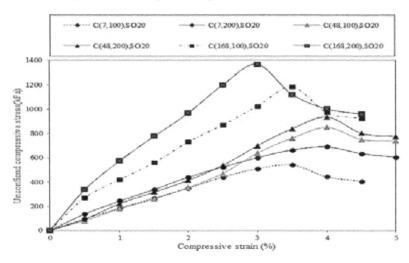

*Figure 10.23* Unconfined compressive strength of carbonated olivine treated soil under carbonation pressures of 100 and 200 kPa after 7, 48 and 168 hours

Source: Fasihnikoutalab (2016)

(a)  100kPa $CO_2$ pressure after 168 hours carbonation time

*Figure 10.24* SEM image of carbonated 20% of olivine treated soil at different $CO_2$ pressures

Source: Fasihnikoutalab (2016)

(b)  200kPa $CO_2$ pressure after 168 hours carbonation time

*Figure 10.24* (Continued)

XRD analyses (Figure 10.25) identified the magnesium based phases of dypingite, hydro-magnesite, and nesquehonite as being responsible for strength development following carbonation with pressure of 200 kPa at 7, 24 and 168 hours of carbonation time. As can be seen by increasing the carbonation time, the amount of these crystals peak increases and the strength of the treated soil increases according to UCS results.

## 10.6   Biogrouting

The concept of biogrouting is to improve soils based on microbiologically precipitated calcium carbonate, $CaCO_3$. In another word using microorganisms to produce the carbonate, which is also known as biocement. Most works on biological grouting (biogrouting or microbially induced carbonate precipitation, MICP) use urease enzyme producing microorganisms, in particular, the bacterium *Sporosarcina pasteurii* in coarse-grained soil.

The bacteria *S. pasteurii* precipitates $CaCO_3$ by producing urease enzyme. The enzyme hydrolyses urea to $CO_2$ and ammonia, resulting in an increase of the pH and carbonate concentration. The carbonates precipitated within the soil matrix filled up the soil pores and hence increase the soil strength, as illustrated in Figure 10.26.

The feasibility of using biogrouting as a ground improvement technique has been demonstrated in the laboratory using sand column experiments (DeJong et al., 2006, Whiffin et al., 2007). A scale-up experiment by Van Paassen et al., 2009 showed significant cementation at a large distance from the injection points, proving the technical feasibility of biogrouting for ground reinforcement in sandy soil as shown in Figure 10.27. However, the $CaCO_3$ content

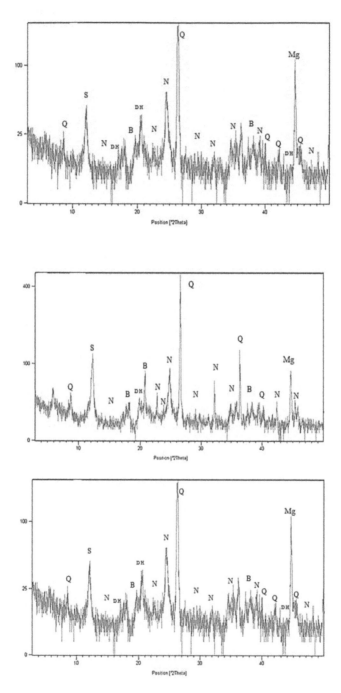

*Figure 10.25* XRD analyses of carbonated olivine-treated soil

Source: Fasihnikoutalab (2016)

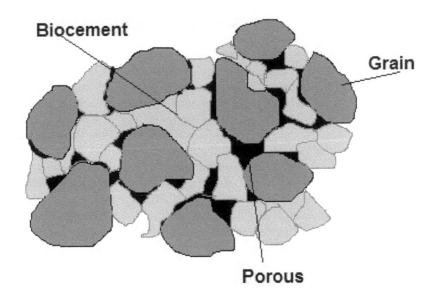

*Figure 10.26* Biocement in porous media of the sandy soil

Source: Whiffin et al. (2007)

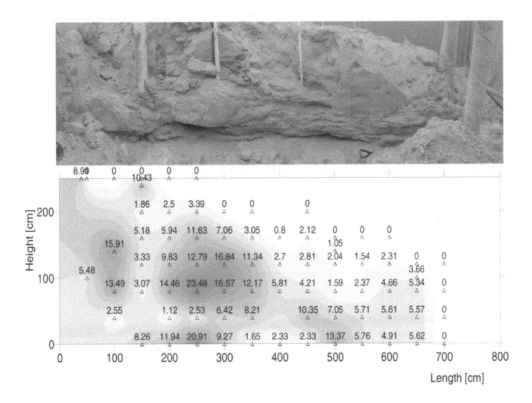

*Figure 10.27* Cross-section along the longitudinal centerline through the center injection and extraction well of a large-scale biogrouting experiment showing CaCO$_3$ concentration

Source: Van Paassen et al. (2009)

and consequent geotechnical parameters were not as homogeneously distributed as desired, both on a macro and micro scale.

Recent work done at Universiti Putra Malaysia attempted at using this technique to stabilise fine-grained soil and transport the bacteria using electrokinetic (EK) stabilisation. Electrokinetic (EK) stabilisation is a ground improvement method in which stabilising agents are induced into soil under direct current (Alshawabkeh and Sheahan, 2003; Asavadorndeja and Glawe, 2005; Barker et al., 2004; Afshin, 2010). The movement of stabilising agents into the soil masses is governed by the principles of electrokinetics, while the mechanisms of stabilisation can be explained by the principles of chemical stabilisation. When ions are used as stabilising agents, the ions migrate into soils through processes of electro-migration and osmotic convection. These ions improve the soil strength by three mechanisms: ion replacement, mineralisation and precipitation of bacteria cells in the pore fluid.

The size of *S. pasteurii* is about 2–3 μm in culture media. Since fine soils have a porous media of 2–3 μm, it is theoretically possible to transport this strain into the fine soils.

In the experiment done, a blend of calcium chloride solution (2M) and urea solution (1M) was injected into the anode chamber over a period of three days. Then, a live culture of *S. pasteurii* bacteria with an optical density of around 0.5 (106 to 108 cells ml-1) was added into the cathode chamber over a treatment time of two days. The bacteria were then injected into the cathode chamber during the polarity reversal for two days. A voltage gradient of 60 mV was applied between the anode and cathode. Shear strengths were measured using vane shear cross the soil sample (from the cathode to the anode) after 7 days of treatment. Water content and percentage of $CaCO_3$ were also measured at each section. The $CaCO_3$ percentage was determined by using an acid washing technique (Mortensen et al., 2011). Figure 10.28 shows the experimental set-up.

A significant increase in soil undrained shear strength after treatment with the greatest improvement near the cathode was found as shown in Figure 10.29.

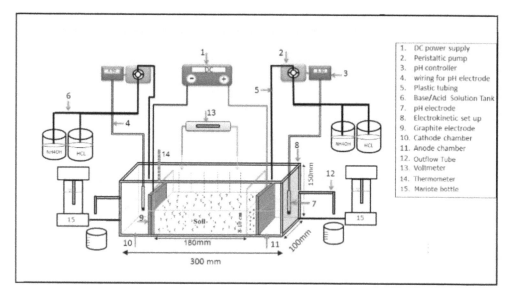

*Figure 10.28* Experimental setup for EK treatment of biogrouting

Source: Kekhya (2013)

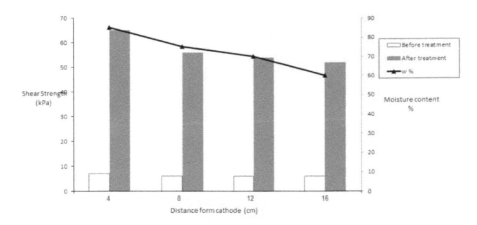

*Figure 10.29* The undrained shear strength and moisture content before and after EK treatment

Source: Kekhya (2013)

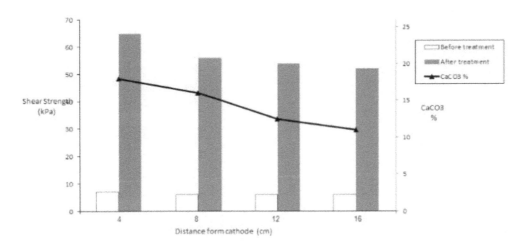

*Figure 10.30* The percentages of CaCO₃ deposits across the soil sample relate to soil shear strength

Source: Kekhya (2013)

The percentage of $CaCO_3$ deposits across the soil sample ranged from 10% to 18% as shown in Figure 10.30. Scanning electron microscopy provides microstructure images of the $CaCO_3$ crystals between the soil particles as shown in Figure 10.31. The energy-dispersive X-ray analysis showed a high concentration of calcium and carbonate in the soil sample after the treatment as shown in Figure 10.32.

This study has demonstrated that it is possible to electrokinetically inject the bacterial products ($CO_3^{-2}$) to enhance the mechanical properties of soils with low permeability.

*Figure 10.31* SEM image of CaCO₃ crystals and clay minerals
Source: Kekhya (2013)

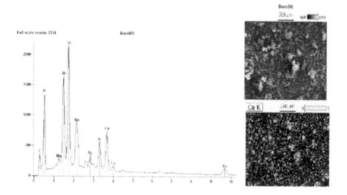

*Figure 10.32* EDX analyses of the soil sample detect calcium concentration
Source: Kekhya (2013)

## Case study 1 (Electrokinetics)

### EKG stabilisation of a railway embankment

Electrokinetic geosynthetic (EKG) technology has been used to successfully stabilise a failing clay embankment in London, resulting in a 26% cost reduction and a 47% reduction in carbon footprint over conventional methods.

There are 20,000 km of earth structures (cuttings and embankments) on the UK highway and rail networks. Few were built to modern geotechnical engineering standards. The ongoing maintenance and remediation that these structures now require has become a major engineering issue for many UK infrastructure owners.

Toe weighting and/or slope regrading is commonly used to tackle the problem, but these do not address the problem of shrink-swell or pore water pressure changes and typically delay failure rather than prevent it. In addition, these methods can consume large quantities of primary aggregate and energy and are becoming less environmentally and economically viable.

Network Rail identified EKG ground treatment as a novel slope treatment method which could:

- Stabilise the slope
- Require only modest access owing to the absence large plant
- Involve low relative energy consumption
- Reduce cost.

A trial was conducted on a 22 m stretch of a 9 m high Victorian embankment. The embankment had been constructed by end tipping a mixture of weathered London clay and other material such as brick and stone fragments onto underlying alluvium and terrace gravels (Figure 10.33). An assessment of the embankment identified several sections as unstable. Inclinometer readings indicated a slip surface at approximately 2.5 m depth, which could either be a shallow translational slide or a deeper circular failure. Stability calculations indicated a factor of safety (FoS) for the slope of only 1.

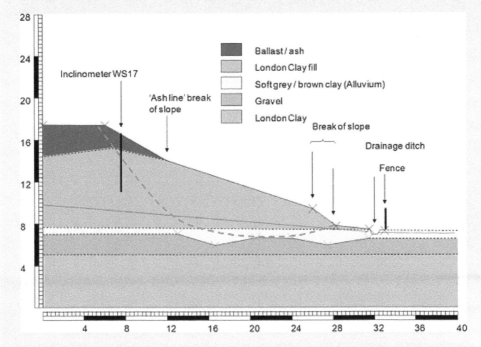

*Figure 10.33* Cross-section of the slope showing the stratigraphy and postulated failure planes (copyright of Electrokinetic Limited, http://electrokinetic.co.uk)

### EKG treatment

EKG treatment was designed to accommodate either of the identified failure mechanisms. The treatment was based around an array of EKG electrodes installed at 2 m centers in the form of tessellating hexagonal cells, with the hexagon being defined by anode stations and a central cathode (Figure 10.34).

Upon application of a DC potential (60–80V) electro-osmosis forced water to flow from the soil adjacent to the anodes to the cathodes. The treatment took only six weeks and resulted in:

- Dewatering from the cathodes >25 times that from control drains
- A reduction in plasticity and shrinkage characteristics
- An increase in groundwater temperature from 10°C to 20°C
- A modest DC power consumption of only 11.5 kWhrs/m³ of soil treated
- Improvements in shear strength parameters ($c$ and  )
- A 263% improvement in the bond strength of the anodes acting as nails
- A cessation of slope movement.

Following EKG treatment the anodes have been retained as permanent soil nails and the horizontal cathodes retained to act as permanent drainage.

Slope stability analyses were undertaken pre- and post-treatment. The analytical results are shown in Table 10.3.

### Longevity of treatment

The use of EKG to stabilise slopes is a long-term solution because:

*Figure 10.34* An array of EKG electrodes installed (copyright of Electrokinetic Limited, http:// electrokinetic.co.uk)

Table 10.3 Results of stability analysis

| Analysis | Reinforcement | FoS(ULS) |
|---|---|---|
| Pre-treatment (TGP data) | No | 0.81 |
| Pre-EKG treatment | No | 096 |
| Post-EKG treatment | No | 1.47 |
| Post-EKG treatment | Yes | 1 71 |

- Soft weak embankment materials consolidate and improve in shear strength with EKG treatment. This consolidation is permanent.
- Additionally, EKG treatment works best on these soft materials, which are critical to the stability, and in this way the treatment can be considered "self-selecting."
- Modifications in soil clay chemistry such as cementation and plasticity occur under conditions induced by electro-osmotic flow. Given the fine-grained nature and very low transmissivity of the soil, the probability of the reversal of these changes is negligible and hence the effects are considered permanent.
- Enhancement of soil/reinforcement bond is a long-term effect.
- Passive drainage (deactivated cathodes) is retained in the slope.

## Costs

A cost analysis comparing slope stabilisation using the EKG method with the lowest cost alternative of gabion baskets and slope slackening indicated that the EKG treatment produced total project cost savings of 26%.

## Carbon Footprint

A carbon footprint comparison of the EKG and conventional treatment options showed 47% lower emissions by using EKG.

## Induced currents

Issues have been raised regarding the possibility of "stray" currents. For clarification, this term is used to denote electric currents which do not flow where intended and are caused by two mechanisms:

- Direct conduction
- Induced currents.

An analysis of the EKG treatment indicated that such currents are negligible.

**Benefits of EKG treatment**

In summary, the benefits of EKG treatment include:

- Effective method for slope stabilisation
- Reduced cost
- Reduced access requirements for labor and plant and materials
- Reduced health and safety risk
- Rapid deployment and low labor requirements
- The treatment can proceed whilst maintaining the railway in service (as occurred during the trial)
- Long term drainage of the slope can be provided for by the filtration and drainage functions of the EKGs in the passive mode
- Sustainability benefits including reduced carbon footprint and elimination of the use of primary aggregates.

Source: http://electrokinetic.co.uk/slopestabilisation.htm

## References

Alsafi, S. (2017) *Gypseous Soil Stabilization by Alkaline Activation Method.* PhD (unpublished), Universiti Putra Malaysia.

Alshawabkeh, A. N. & Sheahan, T. C. (2003) Soft soil stabilisation by ionic injection under electric fields. *Proceedings of the ICE-Ground Improvement,* 7(4), 177–185.

Asavadorndeja, P. & Glawe, U. (2005) Electrokinetic strengthening of soft clay using the anode depolarization method. *Bulletin of Engineering Geology and the Environment,* 64(3), 237–245.

Bakharev, T., Sanjayan, J. G. & Cheng, Y. (1999) Alkali activation of Australian slag cements. *Cement and Concrete Research,* 29(1), 113–120.

Barker, J. E., Rogers, C.D.F., Boardman, D. I. & Peterson, J. (2004) Electrokinetic stabilisation: An overview and case study. *Proceedings of the ICE-Ground Improvement,* 8(2), 47–58.

Basha, E., Hashim, R., Mahmud, H. & Muntohar, A. (2005) Stabilization of residual soil with rice husk ash and cement. *Construction and Building Materials,* 19(6), 448–453.

Bell, F. (1996) Lime stabilization of clay minerals and soils. *Engineering Geology,* 42(4), 223–237.

Correia, A. A., Oliveira, P.J.V. & Custódio, D. G. (2015) Effect of polypropylene fibres on the compressive and tensile strength of a soft soil, artificially stabilised with binders. *Geotextiles and Geomembranes,* 43(2), 97–106.

Davidovits, J. (1979) SPE PATEC '79. Society of Plastic Engineering. Brookfield Center, USA.

Davidovits, J. (1988) Geopolymer chemistry and properties. Paper presented at the *Proceeding of 1st European Conference on Soft Mineralurgy (Geopolymere '88).* Compiegne, Paris.

De Jong, J. T., Fritzges, M. B. & Nüsslein, K. (2006) Microbially induced cementation to control sand response to undrained shear. *Journal of Geotechnical and Geoenvironmental Engineering,* 132(11), 1381–1392.

Duxson, P., Fernández-Jiménez, A., Provis, J. L., Lukey, G. C., Palomo, A. & Van Deventer, J. (2007) Geopolymer technology: The current state of the art. *Journal of Materials Science,* 42(9), 2917–2933.

Fasihnikoutalab, M. H. (2016) *Utilization of Olivine for Soil Stabilization.* PhD Thesis (unpublished), Universiti Putra Malaysia.

Fasihnikoutalab, M. H., Asadi, A., Unluer, C., Huat, B. K., Ball, R. J. & Pourakbar, S. (2017) Utilization of alkali-activated olivine in soil stabilization and the effect of carbonation on unconfined compressive strength and microstructure. *Journal of Materials in Civil Engineering,* 29(6), 601–613.

Fasihnikoutalab, M. H., Pourakbar, S., Ball, R. J. & Huat, B. K. (2017) The effect of olivine content and curing time on the strength of treated soil in presence of potassium Hydroxide. *International Journal of Geosynthetics and Ground Engineering*, 3(2), 1–10.

Fawer, M., Concannon, M. & Rieber, W. (1999) Life cycle inventories for the production of sodium silicates. *The International Journal of Life Cycle Assessment*, 4(4), 207–212.

FHWA. (2003). Geotechnical Engineering Circular No. 7 Soil Nail Walls Publication FHWA-IF-03–017. Federal Highway Administration. US Department of Transportation.

Fox, P. J. & Edil, T. B. (1994). Temperature induced one-dimensional creep of peat. In: Den Haan et al. (eds) *Proceedings of Conference on Advances in Understanding and Modeling the Mechanical Behavior of Peat*. Balkema, Delft, The Netherlands. pp. 27–34.

Glukhovsky, V. (1965) *Soil Silicates: Their Properties, Technology and Manufacturing and Fields of Application*. PHD Thesis, Civil Engineering Institute of Kiev.

Habert, G., D'Espinose de Lacaillerie, J. & Roussel, N. (2011) An environmental evaluation of geo-polymer based concrete production: Reviewing current research trends. *Journal of Cleaner Production*, 19(11), 1229–1238.

Harichane, K., Ghrici, M., Kenai, S. & Grine, K. (2011) Use of natural pozzolana and lime for stabilization of cohesive soils. *Geotechnical and Geological Engineering*, 29(5), 759–769.

Janz, M. & Johansson, S. (2002) The function of different binding agents in deep stabilization. *Swedish Deep Stabilization Research Centre, Report*, 9, 1–35.

Kamon, M. & Nontananandh, S. (1991) Combining industrial wastes with lime for soil stabilization. *Journal of Geotechnical Engineering*, 117(1), 1–17.

Kekhya, H. (2013) *Electro-Biogrouting Stabilization of Kaolin Soil*. PhD Thesis (unpublished), Universiit Putra Malaysia.

Khale, D. & Chaudhary, R. (2007) Mechanism of geopolymerization and factors influencing its development: A review. *Journal of Materials Science*, 42(3), 729–746.

Lizzi, F. (1978) *Reticulated Root Piles to Correct Landslides*. ASCE Convention, Chicago.

Lorenzo, G. A. & Bergado, D. T. (2006) Fundamental characteristics of cement-admixed clay in deep mixing. *Journal of Materials in Civil Engineering*, 18(2), 161–174. doi: 10.1061/(ASCE)0899–1561(2006)

McLellan, B. C., Williams, R. P., Lay, J., Van Riessen, A. & Corder, G. D. (2011) Costs and carbon emissions for geopolymer pastes in comparison to ordinary Portland cement. *Journal of Cleaner Production*, 19(9), 1080–1090.

Miura, N., Horpibulsuk, S. & Nagaraj, T. (2001) Engineering behavior of cement stabilized clay at high water content. *Soils and Foundations*, 41(5), 33–45.

Mortensen, B. M., Haber, M. J., DeJong, J. T., Caslake, L. F. & Nelson, D. C. (2011) Effects of environmental factors on microbial induced calcium carbonate precipitation. *Journal of Applied Microbiology*, 111(2), 338–349.

Nematollahi, B., Sanjayan, J. & Shaikh, F.U.A. (2014) Comparative deflection hardening behavior of short fiber reinforced geopolymer composites. *Construction and Building Materials*, 70, 54–64.

Nixon, J. E. & Mongestern, N. K. (1973). The residual stress in thawing soils. *Canadian Geotechnical Journal*, 10(4), 571–580.

Ortigao, J.A.R. & Palmeira, E. M. (2004) Chapter 13. In *Handbook of Slope Stabilisation*. Springer, Berlin, Heidelberg. pp. 355–388.

Pourakbar, S. (2015) *The Use of Alkali-Activated Palm Oil Fuel Ash (POFA) Reinforced by Microfibers in Deep Mixing Technique*. PhD Thesis (unpublished), Universiti Putra Malaysia.

Pourakbar, S., Asadi, A., Huat, B. & Fasihnikoutalab, M. H. (2015) Soil stabilization with alkali-activated agro-waste. *Environmental Geotechnics*, 2(6), 359–370.

Pourakbar, S., Asadi, A., Huat, B. B., Cristelo, N. & Fasihnikoutalab, M. H. (2016) Application of Alkali-activated agro-waste reinforced with wollastonite fibers in soil stabilization. *Journal of Materials in Civil Engineering*, 29(2), 401–413.

Pourakbar, S. & Huat, B. K. (2016) A review of alternatives traditional cementitious binders for engineering improvement of soils. *International Journal of Geotechnical Engineering*, 11(2), 206–216.

Prusinski, J. R. & Bhattacharja, S. (1999) Effectiveness of Portland cement and lime in stabilizing clay soils: Transportation research record. *Journal of the Transportation Research Board*, 1652(1), 215–227.

Purdon, A. (1940) The action of alkalis on blast-furnace slag. *Journal of the Society of Chemical Industry*, 59, 191–202.

Sargent, P., Hughes, P. N., Rouainia, M. & White, M. L. (2013) The use of alkali activated waste binders in enhancing the mechanical properties and durability of soft alluvial soils. *Engineering Geology*, 152(1), 96–108.

Schlosser, F. (1991) Discussion: The multicriteria theory in soil nailing. *Ground Engineering*, November. pp. 30–33.

Shaikh, F.U.A. (2013) Review of mechanical properties of short fibre reinforced geopolymer composites. *Construction and Building Materials*, 43, 37–49.

Shi, C., Jiménez, A. F. & Palomo, A. (2011) New cements for the 21st century: The pursuit of an alternative to Portland cement. *Cement and Concrete Research*, 41(7), 750–763.

Sukontasukkul, P. & Jamsawang, P. (2012) Use of steel and polypropylene fibers to improve flexural performance of deep soil–cement column. *Construction and Building Materials*, 29, 201–205.

Van Paassen, L. A., Harkes, M. P., Van Zwieten, G. A., Van der Zon, W. H., Van der Star, W.R.L. & Van Loosdrecht, M.C.M. (2009) Scale up of BioGrout: A biological ground reinforcement method. In *Proceedings of the 17th International Conference on Soil Mechanics and Geotechnical Engineering*. IOS Press BV, Amsterdam, Netherlands, pp. 2328–2333.

Vyalov et al. (1978). Stability of mine in Frozen soils. *Proceedings International Symposium on Ground Freezing*. Elsevier, Bochum, 1, pp. 207–215.

Weil, M., Dombrowski, K. & Buchawald, A. (2009) Life-cycle analysis of geopolymers: Geopolymers, structure, processing, properties and applications. *Woodhead Publishing Ltd*, 5(3), 194–210.

Weng, L. & Sagoe-Crentsil, K. (2007) Dissolution processes, hydrolysis and condensation reactions during geopolymer synthesis: Part I—Low Si/Al ratio systems. *Journal of Materials Science*, 42(9), 2997–3006.

Whiffin, V. S., van Paassen, L. A. & Harkes, M. P. (2007) Microbial carbonate precipitation as a soil improvement technique. *Geomicrobiology Journal*, 24(5), 417–423.

Yip, C. K., Provis, J. L., Lukey, G. C. & van Deventer, J. S. (2008) Carbonate mineral addition to metakaolin-based geopolymers. *Cement and Concrete Composites*, 30(10), 979–985.

Yunsheng, Z., Wei, S., Zongjin, L., Xiangming, Z. & Chungkong, C. (2008) Impact properties of geopolymer based extrudates incorporated with fly ash and PVA short fiber. *Construction and Building Materials*, 22(3), 370–383.

# Chapter 11

# Site investigation, instrumentation, assessment and control

## 11.1 Introduction

The knowledge about the site plays a vital role in the safe and economical development of a site. A thorough investigation of the site is an essential preliminary work to the construction of any civil engineering work. Public building officials may require soil data together with the recommendations of a geotechnical consultant prior to issuance of a building permit. Elimination of the site exploration, which usually ranges from about 0.5%–1% of total construction costs, only to find after construction has started that the foundation must be redesigned is certainly uneconomical and unwise. This is generally recognised, and it is doubtful if any major structure is currently designed without exploration being undertaken.

Soil instrumentation is a complex and rapidly evolving field of study. Geotechnical instrumentation refers to the instruments used to monitor geotechnical projects or sites requiring such monitoring. Geotechnical instrumentation and monitoring are essential for the successful completion of a geotechnical projects. Limited geotechnical instrumentation may be needed for simple projects, but the demands on geotechnical instrumentation and monitoring can be very demanding for critical projects such as tunnels, slopes and excavations next to sensitive structures. The amount of instrumentation used in site investigation depends on the type of investigation being carried out. In practice, the amount of instrumentation used in routine pre-design site investigation is very limited and normally consists only of pore water pressure measuring devices. In the case of investigations for deep excavations in rocks, measurements of in situ stress are also made. In contrast to this, trial construction, the investigation of the safety of existing works, and the investigation of failures to allow the design of remedial works all typically involve considerable and quite variable instrumentation. The main parameters which may require measurement are displacement, strain, stress and force; pressure in the form of pore water pressure will be the most frequent measurement because of the relative importance of this parameter in geotechnical design.

## 11.2 Site investigation

In this chapter we will give an overview on the role of site investigation, techniques used and problem of sampling for laboratory tests. In the test section, we will give outline on types of tests usually performed in the laboratory as well as on site (in situ) for obtaining soil parameters, especially for compressible soils.

The allocation for conducting a site investigation is usually a small fraction of the overall project cost. But its importance is enormous. The main objective of conducting a site investigation is to obtain information related to the foundation condition as well as the availability and suitability of natural construction material such as borrow pit or quarry. The investigation is done in the office as well as in the field that is based on the concept of "learn as you go." Data obtained has to be well organised to clearly show the important characteristics such as the formations and properties of soils and rocks. With this the main objective of carrying out a site investigation can be listed as follow:

1.  To evaluate the general suitability of the site for the proposed job
2.  To enable an adequate and economical design to be prepared
3.  To identify any problems that may arise during the construction and propose their solutions.

Meanwhile the specific objective of site investigation, in particular for soil, may involve the following:

1.  Location of sequence, thickness and area of each soil stratum, including description and classification of soil and its structures, and its stratification in the undisturbed condition. Geological properties and other important parameters such as folds, joints, fissures, aggregations, minerals and chemical content.
2.  Type of bedrock including location, sequence thickness, area, altitude and depth of weathering in each rock stratum.
3.  Groundwater characteristics, whether the water table is perched or normal, depth and pressure in the artesian zone and quantity of soluble salts or other mineral presence.
4.  Description of soil characteristics according to the following procedures:

    •   To describe and identify in situ soil and determine water content and basic index properties of the soil. Physical properties are estimated based on identification and results of laboratory index tests.
    •   Indirect tests conducted on site, such as geological interpretation or geophysics using results of other test to obtain appropriate correlations. By observing behavior of other structures built on the same site or other sites of similar soils. Perform field tests such as the standard penetration test, pile loading test, permeability test and vane shear tests. Disturbed and undisturbed samples are collected for laboratory classification tests, soil characteristics such strength parameter, permeability, compressibility and others.

### 11.2.1   Stages of site investigation

Stages in site investigation can generally be listed as follows.

#### 11.2.1.1   Stage 1: Preliminary evaluation of site and ground condition (desk study)

Before any field investigation is carried out, the initial planning for a particular project is done by using available information such as:

•   Topography map
•   Geology map
•   Soil map
•   Geological and geotechnical studies of area of the same geological formation
•   Aerial photo

- Publish record or documents
- Mining history if any, and other engineering information from previous investigation on the same site or from nearby sites.

The main aim of doing this preliminary investigation is so that the engineer would have for his or her perusal a tentative development plan showing type of building and facilities required, size estimate, capacity and general location.

### 11.2.1.2   Stage 2: Preliminary site investigation

The work at this stage has to be done at a minimum cost and time. This preliminary investigation is to obtain a general evaluation of the subsurface condition of the said project, and if necessary also an evaluation aspects of alternative foundation sites.
   Procedures carried out are as follows:

- Visit to site for purpose of making note on land use, history, geological and geotechnical mapping, observing local structures for visible damage, structure type and loading, local services like water pipe, gas pipe, electric and telephone line, accessibility to site, availability of construction material like borrow pit, quarry and local standards on how to prepare and present information.
- Preliminary drilling to collect disturbed and undisturbed samples.
- Static and dynamic penetration tests.
- Preliminary hydrological observation and groundwater condition.
- Simple laboratory tests and in situ test like the vane shear test.
- Geophysical study to obtain general profile of the site.

Size and number of investigation at this stage depends on degree of soil heterogeneity, and of course available funding. For example, for construction of line structure such as road, railway and others, for investigation at this stage, the boreholes could be spaced as follows:

- 1 borehole for every 50–100 m for site on compressible soil of limited area
- 1 borehole for every 200–300 m for site in extensive compressible zone on variable ground surface
- 1 borehole for every 500 m for site on area with homogeneous ground surface.

All boring (drilling) has to be done up to depth of 2–3 m in non-compressible stratum (hard layer with the standard penetration test, $N > 50$, or rock). But in cases where the compressible soil layer is very thick and the applied load is rather small, the depth of the drilling is usually limited to about 2 times the width of the loaded area. For point structure example small building, small pumping station, bridge abutment, normally one borehole is sufficient. But for big structures, at least 4 boreholes (or more) are required.

### 11.2.1.3   Stage 3: Detailed site investigation

Problems identified in stage 2 will be further investigated. The aim is to determine:

- Geometry of soil layer
- The properties of the layer, soil characteristic like its classification, shear strength and compressibility.

- Detail hydrological studies like water level and seasonal variation
- Drainage characteristics.

With the following methods:

- Obtain undisturbed samples for laboratory tests
- Additional penetration tests
- Installation of piezometer
- In situ vane shear test
- In situ permeability test, like pumping test
- Laboratory test for soil classification, shear strength and permeability
- Full scale field trial, field compaction, and pile loading tests, trial embankment and others
- In situ tests like plate loading and pressure meter.

This stage usually begins with conducting fast and cheap tests like the in situ penetration test and vane shear test. Drilling to obtain undisturbed sample from all zones of identified compressible soils is generally done. Drilling is usually done at special location like beneath a bridge abutment, intermediate zone between cut and fill, main structure or high embankment.

## 11.2.2   Sampling techniques

Soil sampling is an important part of a site investigation work. For this various procedures are used.

### 11.2.2.1   Sample quality

Normally soil samples are divided into two classes, that is "undisturbed" and "disturbed" samples. A more accurate classification is given by Rowe (1972), whereby the quality of the sample is divided into five classes based on German classification.

   A disturbed sample is defined as a sample whose structure has been totally or partially disturbed or destroyed, while an undisturbed sample is still in its original state. However, in practice it is not possible for us to obtain a really undisturbed sample. But an effort could be made to minimise the disturbances.

### 11.2.2.2   Main sampling techniques

The main technique for extruding soil sample is as follows:

1.  Mechanical sampling using sampling tube inside borehole or auger hole. For undisturbed samples, thin wall sampler and piston sampler (or core sampler for rock) are used. While for disturbed sample, sampler like the split spoon sampler (see standard penetration test) can be used. The techniques mentioned above are for the inaccessible areas.
2.  For an accessible area, a block sample can be cut by hand from a test hole in order to obtain the undisturbed sample. But with this method, the depth of the sampling is limited to only about 6 m. For deeper depth, the excavation requires side support and water pump

to pump out water if the water table is high. Nevertheless the block sample is much better than the mechanical sample because disturbances to the sample can be minimised.

The choice of technique used determines the quality of the sample required. It must be remembered that the behavior of the whole soil mass is very dependent on the existence of zones of discontinuity or weak zones. Every effort therefore has to be made to note the presence of these zones during boring. Samples obtained have to be representative. Soil structure is another important factor in determining choice of equipment for abstracting sample. For homogeneous, isotropic, fine-grained soil, a 35 mm diameter sampler may be sufficient. But bigger sampler is more representative and gives lower degree of disturbances during sampling. In special cases, a sampler of diameter 150 mm, 200 mm or 300 mm is used.

Main factors that contribute to soil disturbances can be listed as follows:

*   Change of state of stress, example from total stress (while in ground) to zero stress during extraction causing the sample to cavitate)
*   Cyclic stresses during pushing of the sample in to the sampling tube and then withdrawing it
*   Changes in water content due to hydration during preparation of the sample in the laboratory, cavitation and wetting
*   Chemical changes
*   Intermixing and separation of sample matrix.

This sampling effect can be related to the sampler design, method of usage and process of transportation and preparation of the sample.

### 11.2.2.3   Drilling methods

In general the drilling method is limited to the local practice and equipment available. It must however be stated that the drilling work must be done by a competent driller who is reliable and must be under close supervision.

With mechanical sampler, borehole (or auger hole) has first to be drilled to the depth at which the undisturbed sample is to be obtained. For sampling above the water table, the soil may be affected by the wetting. Therefore, for cases of soils like loose and fine sand, dry boring is more suitable. But for fine-grained and saturated soil, wash boring with a cutting bit is more suitable. These soils may not require side support, but normally for cases of soft and saturated soils as well as non-cohesive soil, drilling fluid or casing is needed.

Figure 11.1 shows several types of boring equipment. Table 11.1 summarises the typical drilling method.

### 11.2.3   Sampling equipment

There are several types of sampling equipment that are available.

### 11.2.3.1   Open drive (U4) and thin wall sampler

In general, this sampler (Figure 11.2) is suitable for all types of cohesive soil except if the soil is too hard, cemented or contains gravel. This sampler, however, is not suitable for very wet soil, or the soil, which is too soft as the sample, may fall during the sampling operation.

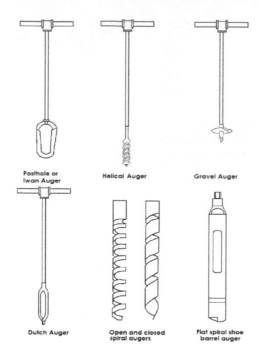

Posthole or
Iwan Auger

Helical Auger

Gravel Auger

Dutch Auger

Open and closed
spiral augers

Flat spiral shoe
barrel auger

(a) Hand-operated auger

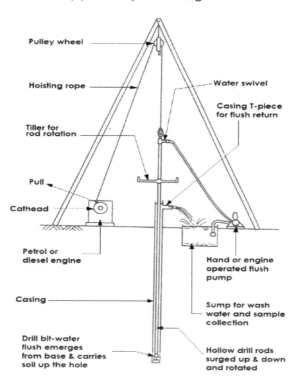

Pulley wheel

Hoisting rope

Water swivel

Casing T-piece
for flush return

Tiller for
rod rotation

Pull

Cathead

Petrol or
diesel engine

Hand or engine
operated flush
pump

Casing

Sump for wash
water and sample
collection

Drill bit-water
flush emerges
from base & carries
soil up the hole

Hollow drill rods
surged up & down
and rotated

(b) Wash boring rig (based on Hvorselev, 1949)

Figure 11.1 Type of drilling

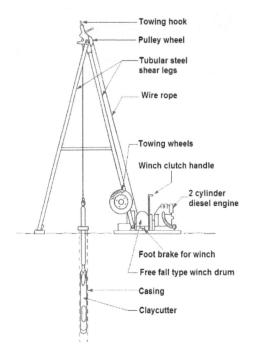

- Towing hook
- Pulley wheel
- Tubular steel shear legs
- Wire rope
- Towing wheels
- Winch clutch handle
- 2 cylinder diesel engine
- Foot brake for winch
- Free fall type winch drum
- Casing
- Claycutter

(c) Light percussion drilling rig (Pilcon Engineering Ltd.)

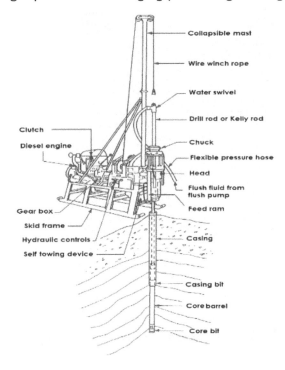

- Collapsible mast
- Wire winch rope
- Water swivel
- Drill rod or Kelly rod
- Clutch
- Diesel engine
- Chuck
- Flexible pressure hose
- Head
- Flush fluid from flush pump
- Feed ram
- Gear box
- Skid frame
- Hydraulic controls
- Self towing device
- Casing
- Casing bit
- Core barrel
- Core bit

(d) Rotary core drilling

*Figure 11.1* (Continued)

Table 11.1 Typical drilling method

| Type of drilling | Borehole support | Disturbed sample | Undisturbed sample |
|---|---|---|---|
| Hand auger | Not available | Able | Able |
| Wash boring | Wash water | In reverse water | Able |
| Mechanical auger | Drilling rod | Disturbed sample | Able |
| Hammer auger | Casing and water in hole | Disturbed sample | Able |
| Rotary auger | Casing and drilling fluid | Only in reverse fluid | Core sample |

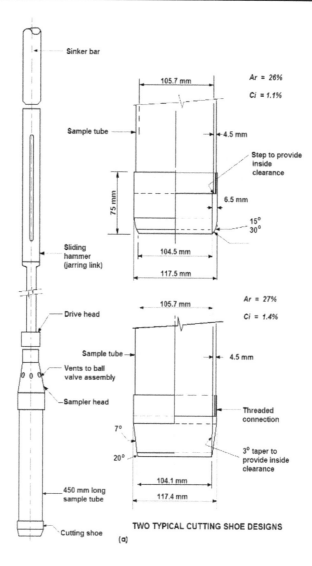

(a) U100 sampler

Figure 11.2 Soil sampler

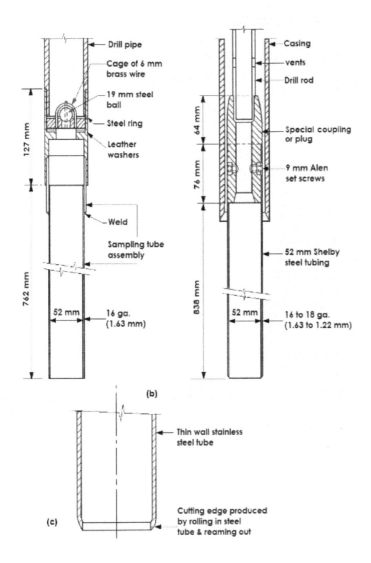

(b) Two thin-wall open-drive samplers

(c) Typical detail of thin wall open-drivers samplers (From Hvorslev, 1949)

*Figure 11.2* (Continued)

Pushing the sampler into the ground in one continuous stroke without rotation does the sampling.

The open drive sampler has a cutting shoe at its end. Sampler size normally used is 100 mm in diameter by 450 mm long. The thin wall sampler on the other hand does not separate cutting shoe but its lower end itself is sharpened. Diameter of the thin wall sampler ranges from 35 mm to 100 mm.

Open drive sampler has an area ratio as high as 30%, while the area ratio of a thin wall sampler is only about 10%. Therefore a thin wall sample enables a better quality of soil sample to be obtained (that is with less degree of disturbances) hence suitable for a soft and rather sensitive soil. Sampler area ratio is defined as:

$$A_r = \frac{d_w^2 - d_c^2}{d_c^2} \qquad (11.1)$$

where

$d_w$ = external diameter of cutting shoe/sampler
$d_c$ = internal diameter of sampler.

### 11.2.3.2   Piston and Bishop sampler

This type of sampler (Figure 11.3) is considered as the most suitable for sampling soft soil that could not be sampled with the thin wall sampler. But this sampler is only recommended for very sensitive soil only as its handling is quite complicated.

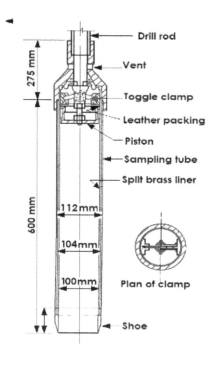

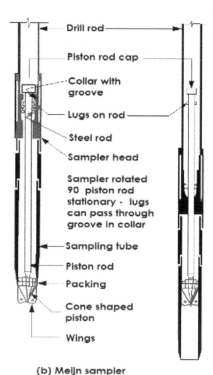

(a) Ehrenberg sampler                (b) Meijn sampler

*Figure 11.3* Two types of piston sampler
Source: Ehrenberg (1933); Huizinga (1944)

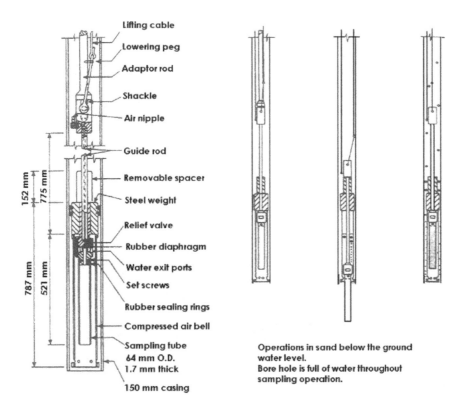

*Figure 11.4* The bishop compressed air sand sampler

Source: Bishop (1948)

### 11.2.3.3  Compressed air sampler

Sampling soil that has no suction such as sand is a special problem. However sampler such as the compressed air sampler could be used for this purpose (Figure 11.4). In this technique the sample is pulled into the diving bell that has suction to retain the sample.

### 11.2.3.4  Begemann and Swedish foil sampler

This type of sampler (Figure 11.5) is suitable for soft soil only. It is particularly good for soil profiling. Foil or plastic stocking that is connected to a piston will be pulled out from magazine as the sampler is pushed into the ground. With this the skin friction between the soil and sampler is avoided, thus the main advantage of this sampler. Long sample of up to 20 m may be obtained using this sampler.

### 11.2.3.5  Peat sampler

As mentioned before, peat is the extreme form of soft soil, hence peat is very difficult to sample in an undisturbed condition. Having said that, there are several samplers that have been developed especially for this difficult material.

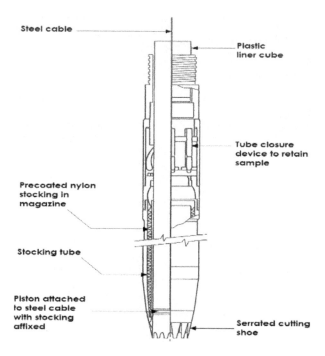

Steel cable

Plastic
liner cube

Tube closure
device to retain
sample

Precoated nylon
stocking in
magazine

Stocking tube

Piston attached
to steel cable
with stocking
affixed

Serrated cutting
shoe

## (a) Begemann sampler

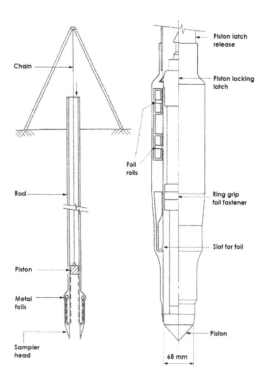

Chain

Rod

Piston

Metal
foils

Sampler
head

Piston latch
release

Piston locking
latch

Foil
rolls

Ring grip
foil fastener

Slot for foil

Piston

68 mm

(b) The principle of operation of the Swedish foil sampler (Clayton et al., 1982)

*Figure 11.5* Begemann and Swedish foil sampler

A 100 mm diameter peat sampler has been developed in Sweden. This sampler consists of a sharp wave-toothed edge mounted on a plastic tube with a driving head at the upper end, Figure 11.6. Samples are taken from the ground surfaces or the bottom of pre-bored holes. After extraction of the sampler, the cutting edge and driving head are removed and the sample in the plastic tube is sealed. Laboratory tests show that samples of fibrous soils taken with this peat sampler have better quality than samples taken with a small diameter piston sampler.

Piston sampler described above may also be used for sampling peat, like the NGI 54 and 95 mm piston sample, while the Japanese recommended a 75 mm piston sampler as standard.

Seaby (2000) has presented a sampler not especially for geotechnical investigation rather for forestry. Despite the purpose of development it can be used to take the undisturbed samples from medium depths. For obtaining peat cores 1 m long or more without compacting them, sampler is described as comprising two halves inserted separately (Figure 11.7). A length of 3.5 mm thick PVC pipe 80 or 110 mm in diameter was halved and the tip of each half was chamfered and provided with pointed, sharp metal blade attached to inside. On this blade was a loop of spring steel wire which, due to its hinging through holes in the outer edges of the blade, was pushed un against its inside circumference during insertion, whereas during extraction a slight downward movement, relative to the sampler, swung it out to help firmly grip the core. For alignment of the two halves during insertion the second half had a wider

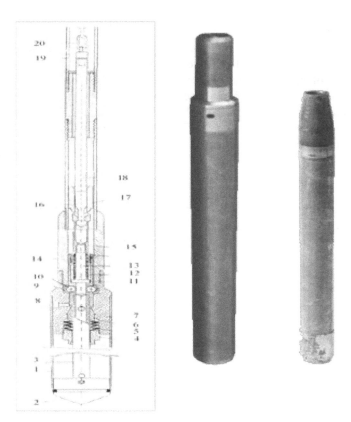

*Figure 11.6* Cross-section of piston sampler

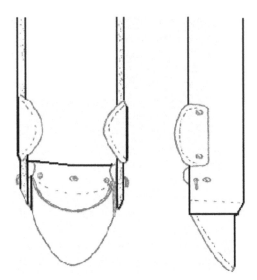

*Figure 11.7* Seaby's peat sampler

cutting blades; one attached to each side near the tip. These closely overlapped the first half externally, acting as guide. Both halves had cross handles near the top, but no aid extraction and reduce the risk of back strain, the halves were bolted together and a series of holes along one length allowed a lever with pointed tip, acting over a fulcrum, to ease the sampler out.

Note that the plan and side view show the end of second half to be driven, the main blade attached to the inside circumference of the plastic tube has two flanges that internally slightly overlap the first half to be driven.

### 11.2.3.6   Block sampling

For special test (such as the centrifuge test) a high quality sample may be required. Sometimes big sample of size up 1 m³ is needed. In such a case block sample may be obtained. The sampling is done by first excavating a hole to the required depth. A block is then cut and a container is then placed around the block. Or the container can also be driven into the ground and then excavating the soil around the container until its base. The container with sample inside is then pushed to one side on top of a steel plate with act as base for transporting the sample to the laboratory. Block sampling is also suitable for undisturbed sampling of shallow peat.

### 11.2.3.7   Split spoon sampler

This sampler comprises two cylindrical tubes with are split along their longitudinal axis as shown in Figure 11.8. The external and internal diameters of the sample are 35 mm and 50 mm, respectively, with an area ratio of about 100%. This sampler is specifically used in the standard penetration test (SPT) for obtaining disturbed samples only.

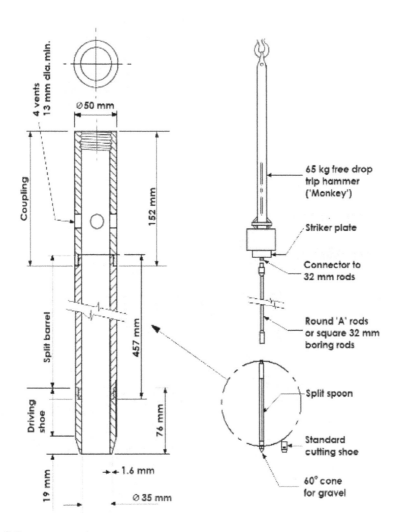

*Figure 11.8* Split spoon sampler

The Mazier corebarrel (Figure 11.9) is a triple-tube swivel type reactor barrel, whose effectiveness relies on the fact that the amount of inner barrel protrusion is controlled by a spring placed in the upper part of the device. The inner barrel contains a brass liner which can be used to transport samples to the laboratory, or for storage. The cutting shoe on the bottom of the inner barrel is substantial making it much less easily damaged than a thin-wall tube, but introducing the problems of disturbance when the high area ratio shoe travels ahead of the corebit. This sampler is suitable for hard friable soil.

### 11.2.3.8  Transporting sample

Vibration can disturb samples, particularly for case of sand and soft sensitive clay. The quality of samples used for laboratory test depends on how the samples are handled after they have

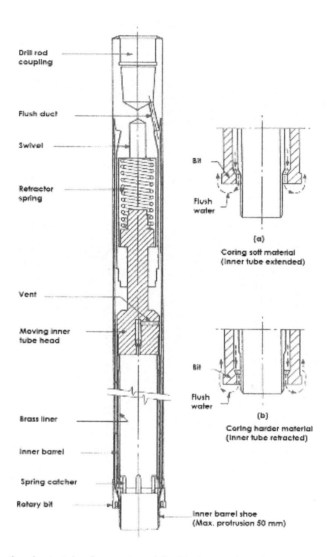

*Figure 11.9* Detail and principle of operation of the Mazier corebarrel

been extruded from the ground. In order to minimise disturbances to the sample, the following general guide could be followed.

- Laboratory tests must be performed as soon as possible after the sampling was made (maximum allowable time period depends on types of soil and laboratory test but usually in the range of a couple of weeks).
- After extraction the sample must be labeled and properly recorded. The records must content information like groundwater level, drilling or sampling method, soil classification, complete log of soil stratum (unless for case of continuous undisturbed sample), method of cleaning the hole prior to the sampling, brief statement about the purpose of

sampling and sketch to show relation of sample with ground surface like base of footing or side slope with dimension and height. For driven sample, record its driving length and length of sample obtained.

• The sample must be sealed as soon as possible on site using wax.
• During transportation to laboratory, the sample must not be exposed to extreme temperature, and then kept in room with constant humidity and temperature.

A representative sample is required for the determination of natural moisture content.

## 11.2.4   Site investigation report

Data obtained at each stage of the site investigation has to be presented in brief in form of bore log. This log includes description of earth stratum, detail of sampling, in situ tests and soil properties. Figure 11.10 shows sample of the bore log. It must be noted here that detail arrangement of the bore logs usually differs, depending on local practice. Besides the above, in situ and laboratory test results can also be shown in form of table, as given in Figure 11.11.

The bore logs should enable a profile of the soil stratum to be made. The profile of soil stratum has to be shown in detail to include the followings:

• Classification (moisture content, particle size, Atterberg's limit) as well as the classification system used, like the Unified Classification System
• Description of undisturbed condition of soil stratum
• Area and extend of stratum with potential instability
• Test parameter like density, permeability, compressibility, and soil shear strength parameters ($c$ and $\phi$)
• Groundwater level
• Chemical content such as organic and sulfate
• Depth of bedrock, thickness of weathering, structural weakness and discontinuity, if any
• Accurate identification of soil during the site investigation is very important for such information is usually as indicator for potential engineering problem that may arise.

Two phases classification procedures are normally used. They are:

• *Preliminary classification*: Preliminary classification is prepared on site based on drilling including bore log. This classification is based on simple observation such as color of the soil, its texture, particle size and organic content, soil stratification, fabric, presence of other soils, fossil, organic materials, foreign debris and so forth. A simple approach to describe each stratum is made by describing the following:

1. Moisture content
2. Color
3. Consistency
4. Soil types
5. Origin
6. Groundwater condition.

## 11.2.4.1   Moisture content

Soil moisture content can be described as dry, damp, very damp and wet.

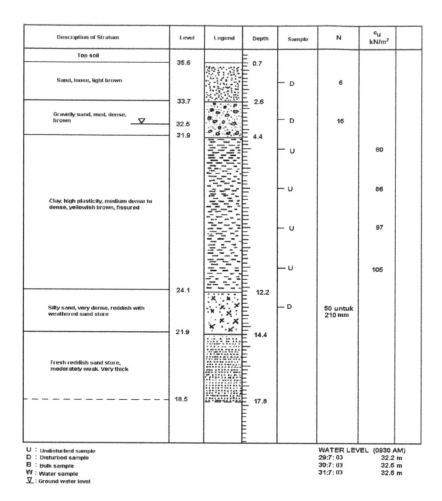

| Description of Stratum | Level | Legend | Depth | Sample | N | $c_u$ kN/m$^2$ |
|---|---|---|---|---|---|---|
| Top soil | 35.6 | | 0.7 | | | |
| Sand, loose, light brown | | | | D | 6 | |
| | 33.7 | | 2.6 | | | |
| Gravelly sand, med. dense, brown | 32.5 | ▽ | | D | 15 | |
| | 31.9 | | 4.4 | | | |
| | | | | U | | 80 |
| | | | | U | | 85 |
| Clay, high plasticity, medium dense to dense, yellowish brown, fissured | | | | | | |
| | | | | U | | 97 |
| | | | | U | | 105 |
| | 24.1 | | 12.2 | | | |
| Silty sand, very dense, reddish with weathered sand stone | | | | D | 50 untuk 210 mm | |
| | 21.9 | | 14.4 | | | |
| Fresh reddish sand stone, moderately weak. Very thick | | | | | | |
| | 18.5 | | 17.8 | | | |

U : Undisturbed sample
D : Disturbed sample
B : Bulk sample
W : Water sample
▽ : Ground water level

WATER LEVEL  (0930 AM)
29:7: 03          32.2 m
30:7: 03          32.5 m
31:7: 03          32.5 m

*Figure 11.10* Sample of a borelog

### 11.2.4.2   Color

Color is a characteristic that can easily be seen. It shows chemical process and mineralogy, especially in relation to iron component. Examples of bright colors are red, purple, orange and yellow. Examples of dark colors are chocolate, olive, green and blue, and examples of intermediate colors are light reddish orange, dark blue green and others. Color changes with moisture content. Therefore it must be recorded while the sample is in undisturbed condition.

### 11.2.4.3   Consistency

Consistency describes hardness or compactness of the soil. Therefore it indicates soil stiffness and strength. The following descriptions are for soil consistency:

Clay: very soft, soft, firm, hard, very hard
Granular soil: very loose, loose, medium dense, dense, very dense.

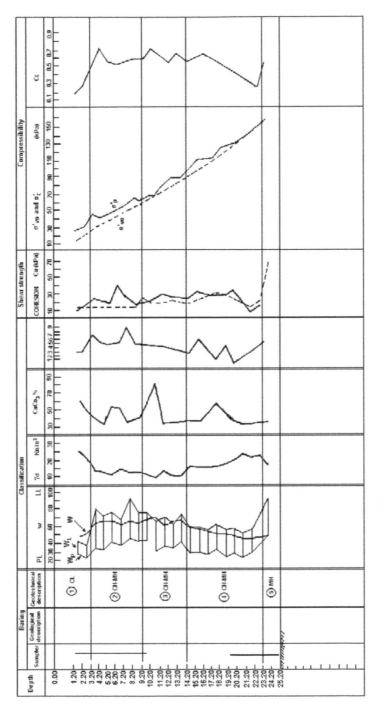

*Figure 11.11* Example of an in situ and laboratory test results

*Table 11.2* Undrained shear strength of clay and granular soil

| Description | Test | $Cu$ (approx.), $kN/m^2$ |
|---|---|---|
| *Clay* | | |
| Very soft | Extrude between figure when squeezed. | < 20 |
| Soft | Easily dent with figure. | 20–40 |
| Quite stiff | Quite difficult to dent with figure, but can be easily peeled. | 40–75 |
| Stiff | Cannot be dented with figure. Difficult to peel. | 75–150 |
| Very stiff | Very stiff and very difficult to peel. | >150 |
| *Granular soil* | | |
| Very loose | Easily excavated with spade. | < 4 |
| Loose | Quite easily excavated with spade or poke with handle. | 4–10 |
| Moderately dense | Difficult to excavate with spade or poke with handle. | 10–30 |
| Dense | Difficult to poke with handle. Require excavation. | 30–50 |
| Very dense | Very difficult to excavate. | > 50 |

Table 11.2 shows approximate values of undrained shear strength (for cohesive soil) and blow counts of the standard penetration test (for granular soil), according to BS 5930: 1981.

### 11.2.4.4    Structure

Structures in this context mean the presence or absent of discontinuity inside the soil mass and not arrangement of particles in soil skeleton. Structure in soil mass plays an important role in controlling soil behavior, especially its shear strength and permeability. Description which refers to its properties are:

> *Bedding*: This refer to visible bedding plane, such as thin layers, thick layers or alternate layers.
> *Discontinuity*: Usually refers to joints or fissures. Undisturbed means no discontinuity. For fissures, record has to be made of the surface and intensity of its discontinuities.
> *Shearing*: The presence of the shear surface is normally indicated by the presence of slickenside (polished surface due to shearing).

### 11.2.4.5    Soil types

Based on particle size, soil can be classified as cobbles, gravel, sand, silt and clay, as shown in Table 11.3.

### 11.2.4.6    Origin

Origin, mode of deposition and geological history can be important in evaluating engineering characteristics of soil. Normally this task is best left to a geologist or at least reference is made to geological memoir.

*Table 11.3* Soil types

| Soil type | Size | (mm) | Description (that can be seen or felt by touching) |
|---|---|---|---|
| Cobble | | >60 | |
| Gravel | Coarse | 20 | Can be seen with coarse eye |
| | Medium | 6 | Particle shape: angular, sub angular, round, flat, elongated |
| | Fine | | |
| | | 2 | |
| Sand | Coarse | | Texture: coarse, smooth, polished |
| | | 0.6 | |
| | Medium | | Grading: well graded, poor graded (uniform), gap graded |
| | | 0.2 | |
| | Fine | | |
| | | 0.06 | |
| Silt | | | Cannot be seen with coarse eye. Gritty feeling. Shows dilatancy when held in hand. Dissolves quickly in water. |
| | | 0.02 | |
| Clay | | | Feels "soapy" when wet. Sticks to figure and dry solely. No dilatancy. |
| Organic soil | | | Content organic matters, mostly plant derivatives |
| Peat | | | Comprises mostly plant remains. Dark brown to black in color. Low density. |

In determining the origin of soil, it is important to differentiate the followings:

*   *Residual soil*: soil formed in situ by weathering process of parent rock.
*   *Deposited soil*: that is material transported by various agents such as wind, water, and gravity and glassier.
*   *Pedocrete*: that is soil that has been cemented or partially replaced by chemical reaction. Laterite, ferricrete, calcrete and silicrete are examples of such soils.

### 11.2.4.7   Groundwater

Groundwater condition at any location is one of the most important factors that determine the ground behavior. Description of ground profile is incomplete unless reference is also made to the groundwater level.

Test holes and tunnels are useful in evaluating groundwater condition, such as level, rate of inflow, seepage influence on consistency and stability of soil sides or slopes. Exploration done using boring and sampling requires standpipes to be installed to determine the groundwater level. Groundwater level as recorded in bore log is not always reliable.

*Definitive definition*: Definitive classification is based on preliminary classification and results of classification tests carried out in the laboratory such as particle size distribution, organic and sulfate content and shrinkage limit.

## 11.3　Investigation report and test methods

The general format for preparation of a site investigation report is as follows.

### 11.3.1　Initial evaluation report

The initial evaluation report consists of:

- Summary of work done and sources of information
- Preliminary assessment of ground and site condition in relation to the project
- Suggestion for next stage of investigation.

### 11.3.2　Preliminary site investigation report

The preliminary site investigation report consists of:

- Summary of work done
- Summary of soil data studied in form of simple cross-section together with estimated characteristics of soil
- Statement of problem that may arise when the project is implemented.
- Suggestions for further action.

### 11.3.3　Detailed site investigation

The detailed site investigation consists of:

- Summary of work done.
- Summary of study, for example cross-section of each section of soil that has been studied, including mechanical properties of layers, average mechanical properties of soil layer for consideration in calculation for stability and settlement.
- Proposed design approach. Additional observation and investigation if needed. Problem that may arise during construction stage and action that has to be taken.

Figure 11.12 shows an example of a ground profile plot.

### 11.3.4　Test methods

There are a number of tests that are normally done in situ and in the laboratory.

### 11.3.5　Laboratory tests

*Shear strength tests*: Tests that are normally carried out in the laboratory to determine shear strength parameter are unconfined compression test, direct shear test and triaxial test. In addition to this, other tests, such as Swedish fall cone and lab vane shear may also carried out.

Bjerrum (1973) considers the suitability of various shear tests in relation to condition along potential slip plane of an embankment as shown in Figure 11.13.

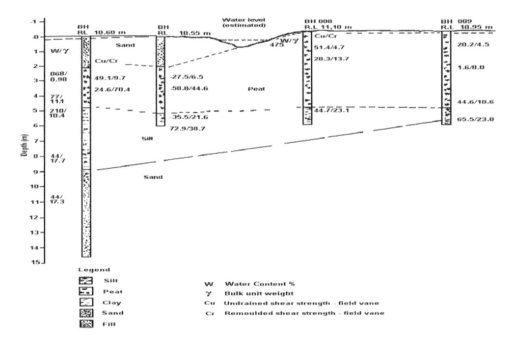

*Figure 11.12* Example of a ground profile plot

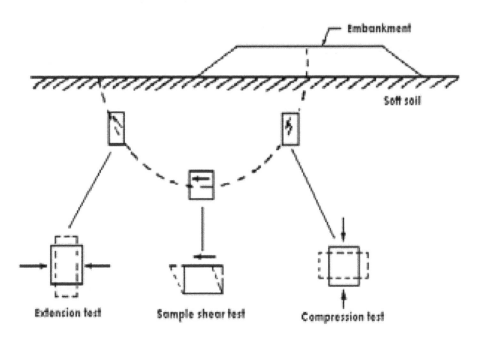

*Figure 11.13* Suitability of various shearing test
Source: After Bjerrum (1973)

Influence of anisotropy on shear strength is also important. In soil like this, orientation of the slip surface has a major influence on the soil shear strength.

For total stress analysis, usually unconfined compression tests and unconsolidated undrained triaxial tests were carried out. In this test, the soil cohesion is assumed to equal to half its stress at failure and soil angle of friction is assumed as zero. Other tests such as Swedish fall cone and vane shear test may also be used.

Consolidated undrained triaxial test (without measurement of pore water pressure) can also be used to evaluate increase in soil shear strength due to consolidation. The result of this test can be used in stability study during construction. However in this case, estimate has to be made of the stresses at various stage of the construction. In some cases such estimate can be difficult to make.

For sensitive soil and soils with micro fabrics and fissures, a large sample up to 250 mm in diameter may be required (Rowe, 1972).

For shear strength parameter base on effective stress, a test such as consolidated undrained triaxial with pore pressure measurement could be performed.

Soil shear strength, however, is not a unique parameter. It depends on a number of parameters such as stress path, test type and soil stress history.

*Consolidation*: Primary consolidation data usually used to predict settlement of compressible soil in usually obtained in the laboratory using the oedometer apparatus. The results obtain correspond to an at rest ($k_o$) condition. In this test, the coefficient of volume compressibility $m_v$ (or $c_c$), and coefficient of consolidation $c_v$ is obtained for vertical flow only. To evaluate coefficient of consolidation in horizontal or radial direction, one way is to orient the sample in such a way that its horizontal axis is in the vertical direction inside an oedometer. Another method for radial drainage, especially for design of a vertical drain such as a sand drain or wick drain, is to have a sample with its central core, through which drainage will occur, filled with sand.

The oedometer test is widely used because it is easy to use. But this test is also known to be quite unreliable in predicting rate of settlement especially for case of soil with micro fabrics or fissures.

Rowe (1972) proposed the use of bigger diameter sample that is 250 mm in diameter by 50 mm thick to overcome this problem. As an alternative in situ permeability test can be carried out.

In determining the consolidation parameter for soil with high organic content such as peat, the use of normal load increment can be misleading. With this a single stage load increment is more suitable. But this requires a number of identical samples to be tested at different loads. Another way of testing peat and high organic soil is to make use of laboratory permeability test together with coefficient of volume compressibility from normal consolidation test. This procedure proves to give a good estimate for this case.

The usual procedure for determining the coefficient of secondary compression ($c_\alpha$) is also by using the oedometer apparatus. That is the after end of the first phase of consolidation, load is maintained over an adequate length of time to ensure a linear relation between strains and log time to occur. The slope of this line is defined as coefficient of secondary compression, $c_\alpha$. Typical value of $c_\alpha$ may range from 0.002 for inorganic clay with low moisture content (20%) to 0.1 for peat with moisture content of 1000%.

Although much research has been done in this area, the mechanism of secondary compression however is still not quite understood.

*In situ tests*: In recent years, the importance of the in situ test has been increasingly recognised. This is because these tests are cheap and enable quick evaluation of general characteristics of extensive soil deposit to be made. Sometimes it is a practical method to obtain reliable data.

In certain situation, subsoil properties, which are too complex and variable, can bring major problem to normal testing and analysis method. In cases like this, an actual loading test or in situ test may be the only approach available for design.

In situ tests has a number of advantageous compared with laboratory test as follows:

> Sample disturbance during sampling, transportation and preparation of sample can be avoided. Furthermore changes in soil properties due to temperature and moisture change can be minimised. Soft sensitive soils needs piston sampler, but this sampler is expensive and requires a long time to operate.

1. In situ test is more representative about soil structure and fabrics. Gravelly soil, for example, is difficult to sample
2. Effect of sample size can be avoided.

However there are also disadvantages of the in situ test. Flow of water from the soil for example cannot be controlled.

In situ tests that are normally performed are:

- Penetration test – SPT and cone penetrometer
- In situ vane shear
- In situ permeability test
- Pressure meter test
- JKR probe.

*Penetration test:*

There are two main type of penetration test, that is:

1. Standard penetration test (SPT)
2. Cone penetration test (CPT).

*Standard Penetration Test (SPT)*: This test measure resistance by recording number of blows of a standard hammer of weight 140 lbs. falling from a height of 76 mm, to drive a split spoon sampler over a depth of 300 mm into the ground. The equipment used is shown in Figure 11.8. The soil sample obtained is categorised as disturbed sample and is usually used for classification tests only. This test, however, is the only reliable test to determine parameters of granular soils.

The results of the standard penetration test can be empirically related to the relative density, bearing capacity factors and angle of frictions of granular soil, as shown in Figure 11.13. For cohesive soil, Terzaghi and Peck (1967) propose the following relation between SPT and unconfined compression (see Table 11.4).

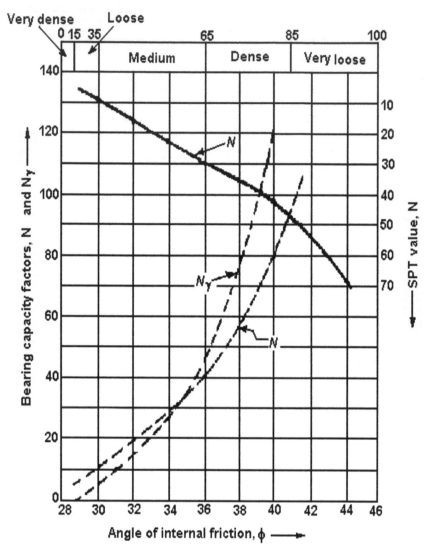

*Figure 11.14* Correlation between N, φ and bearing capacity factors ($N_c$ and $N_g$)

Source: Peck et al. (1974)

*Table 11.4* Relation between N and unconfined compression, $q_u$

| N | Consistency | Unconfined compression strength (kPa) |
|---|---|---|
| 2 | Very soft | 25 |
| 2–4 | Soft | 25–50 |
| 4–8 | Medium | 50–100 |
| 8–15 | Stiff | 100–200 |
| 15–30 | Very stiff | 200–400 |
| >30 | Hard | 400–800 |

### 11.3.6 Cone penetration test (CPT and CPTu)

This test was invented in the 1930s and is mainly used to determine properties of soft to medium soils such as clays and sands. However, with large, rugged modern-day cones, it is now possible to carry out CPTs in dense soils. The test is performed by driving a 60° cone (typically with a cross-sectional area of 10 cm$^2$, but other areas are also used) into the soil at rate of 2 ($\pm$0.5) cm/s. No borehole is required. Begemann (1965) improved the functioning of the earlier cone by adding in friction jacket with which the skin friction is also measured in addition to the cone resistance. This Begemann cone and its simpler sibling (without friction measurement) are known as mechanical cones. Figure 11.15 shows an illustration of the Begemann cone.

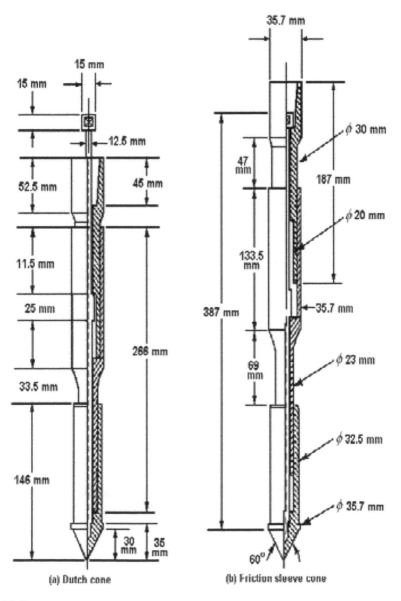

*Figure 11.15* Begemann cone

Nowadays many of the cones come with electric load cells and are equipped with an additional sensor to measure the pore pressure. This is known as CPTu. The test is a special type of cone penetration test (CPT) which allows additional measurement of excess pore pressure generated during the penetration. Indeed, the "u" in CPTu represents the porewater pressure. Due to its efficiency and precision, the CPTu is becoming one of the commonly used in situ testing methods in geotechnical investigation worldwide. Figure 11.16 shows an example of a CPTu equipment.

Compared to mechanical CPT, the electrical CPTu has been significantly improved in terms of precision and acquired data. The data can be continuously acquired at much smaller intervals compared to mechanical CPT.

The cone penetration test is fast and very cost effective to perform. Besides determining soil type a variety of soil parameters can be derived from the basic parameters cone resistance ($q_c$), local sleeve friction ($f_s$) and pore pressure (typically $u_2$).

Undrained shear strength of soil, $S_u$, based on the mechanical CPT, may be determined from cone resistance, $q_c$, using the following equation:

$$S_u = q_c \:/\: factor \tag{11.2}$$

where $q_c$ is the cone resistance. Table 11.5 summarises the empirical cone factor values.

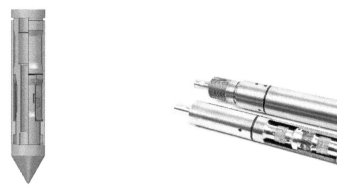

Figure 11.16 CPTu equipment
Source: Courtesy of Geomil Asia Sdn Bhd

Table 11.5 Factor for determination of undrained shear strength

| Country | Factor |
|---|---|
| Belgium | 10–20 |
| France | $(q_c - s_{vo})/S_u = 12$ if $q_c - s_{vo} < 600$ kPa |
| | $(q_c - s_{vo})/S_u = 30$ if $q_c - s_{vo} > 600$ kPa |
| Greece | 15–18 |
| Italy | $15 < (q_c - s_{vo})/S_u < 25$ |
| Netherlands | 15 |

*$s_{vo}$ is overburden pressure

However, the undrained shear strength, *Su*, may vary since the undrained response of soil depends on the direction of loading, soil anisotropy, strain rate, and stress history (Robertson and Capal, 2015). They suggested the following relationships for $S_u$ for the CPTu:

$$S_u = (q_t - \sigma_{vo}) / N_{kt} \tag{11.3}$$

where $q_t$ is a corrected cone resistance $(q_t = q_c + u_2(1 - a))$, $u_2$ is pore pressure when measured just behind the cone and $a$ is the net area ratio determined from laboratory calibration with a typical value between 0.70 and 0.85. In sandy soils, $q_c = q_t$. $N_{kt}$ is a factor which typically varies from 10 to 18, with 14 as an average. In very sensitive fine-grained soil, $N_{kt}$ can be as low as 6.

A chart for soil classification based on CPT data, first proposed by Robertson et al., (1986), later updated in by Robertson in 2010 in its dimensionless form is shown in Figure 11.17. This chart uses the basic CPT parameters of cone resistance, $q_c$, and friction ratio, $R_f$. The chart is global in nature and can provide reasonable predictions of soil type for CPT

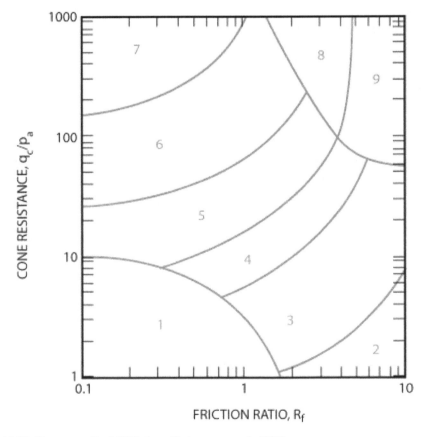

*Figure 11.17* Non-normalised CPT chart (Robertson et al., 1986)

Source: https://websites.pmc.ucsc.edu/~afisher/post/dp/Robertson2010_CPT10_SoilClassCPT.pdf

soundings up to about 20 m in depth. Overlap in some zones should be expected and the zones can be modified somewhat based on local experience (Robertson and Capal, 2015).

| Zone | Soil Behavior Type |
|:---:|:---:|
| 1 | Sensitive, fine-grained |
| 2 | Organic soils - clay |
| 3 | Clay - silty clay to clay |
| 4 | Sill - mixtures clayey silt to silty clay |
| 5 | Sand mixtures - silty sand to sandy silt |
| 6 | Sands - clean sand to silty sand |
| 7 | Gravelly sand to dense sand |
| 8 | Very stiff sand to clayey sand* |
| 9 | Very stiff fine-grained* |

* Heavily over consolidated or cemented
$P_a$ = atmospheric pressure = 100 kPa = 1 tsf

Cone resistance, $q_c$, is the force acting on the cone, $Q_c$, divided by the projected area of the cone, $A_c$.

$$q_c = Q_c / A_c \tag{11.4}$$

Friction ratio, $R_f$, is the expressed as a percentage, of the sleeve friction resistance, $f_s$, to the cone resistance, $q_t$, both measured at the same depth.

$$R_f = (f_s / q_t) \times 100\% \tag{11.5}$$

A CPTu can be also used to carry out pore pressure dissipation test. A dissipation test can be performed at any depth by measuring the rate of excess pore water dissipation of excess pore pressure to reach a certain percentage of equilibrium value representing the hydrostatic water pressure at that depth. The test provides useful information about the porewater pressure dissipation of the soil and its potential for liquefaction.

Other soil parameters that can be estimated based on the CPT data includes (Robertson and Capal, 2015):

* Soil unit weight, soil sensitivity, stress history (over consolidation ratio, OCR), in situ stress ratio, $K_o$, relative density
* Friction angle
* Stiffness and modulus, and consolidation characteristics.

### 11.3.7 Vane shear

The vane shear test was invented in Europe in the 1920s. The vane consists of a steel rod having at one end four projecting blades or vanes parallel to the axis, and situated at 90° intervals around the rod (Figure 11.18). The rod is pushed into the soil to the desired depth, or to a depth of 750 mm below the base of the borehole. The vane is then rotated at a rate of 0.1° per second to reach failure in about 3–10 minutes. The value of maximum torque is recorded.

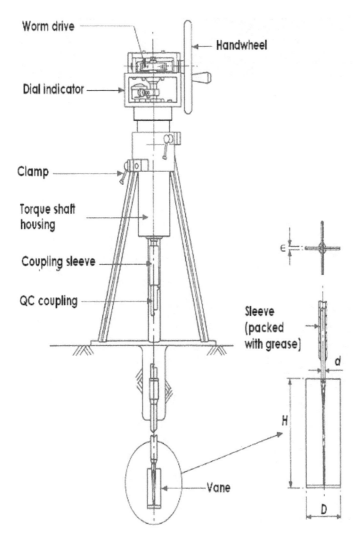

*Figure 11.18* Vane shear

The rotation is then continued to obtain the remolded strength of the soil. With this value of soil sensitivity, ratio of $c_u/c_u$ (remolded) can be calculated.

$$c_u = \frac{T}{\frac{\pi D^2}{2} H \left(1 + \frac{D}{3H}\right)}$$

(11.6)

where

$T$ = applied torque force
$D$ = diameter of vane.

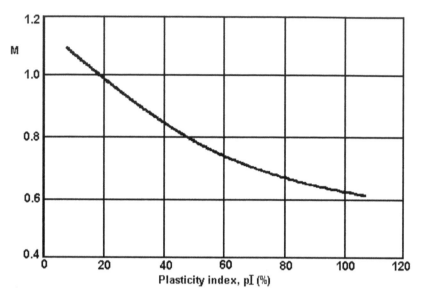

*Figure 11.19* Correction factors for the vane shear test results

Source: Bjerrum (1973)

The ratio of height to diameter ($H/D$) of vane normally used is 2. Figure 11.19 shows correction factors for the vane shear test results.

Main critics of this test are:

1. Failure surface is not of cylindrical shape as assumed
2. Because shear stress mobilised is not uniform, progressive failure may occur in brittle soil
3. Drainage can occur in case of anisotropic (layered) soil
4. Soil at base of borehole where the test is to be performed, is easily disturbed by method of drilling used
5. Bending of connecting rods can occur inside the borehole.

But having said the above, this test is actually very useful especially for very soft soil where sampling could be too difficult and expensive to make.

### 11.3.8 Pressure meter

Pressure meter test is an in situ loading test that is usually done inside a borehole (Figure 11.20). Menard first invents the test in the 1950s. The basic form of a pressure meter comprised a cylindrical cell covered with rubber membrane that can be expanded to the sides of a borehole by pumping in fluid into the cell. Pressure and volume of the fluid is observed continuously. Both the top and bottom cell, above and below the main cell, are also expanded to ensure the main cell is expanding in the radial direction only. Data obtained is plotted in form

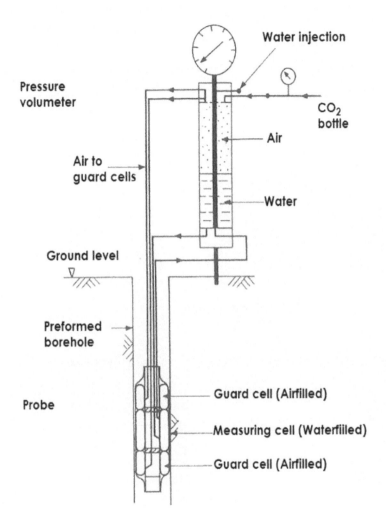

*Figure 11.20* Menard pressure meter
Source: Gibson and Anderson (1961)

of volume change curve versus pressure. From this curve, the shear strength and stress-strain behavior of soil can be estimated.

### 11.3.9 Plate loading test

This test is actually a model footing test (Figure 11.21). Test hole is excavated to the required level of the footing. Steel plate is then placed on the ground inside the hole. Static load is then applied to the plate in a sequence of increments. Magnitude and rate of settlement is recorded. Recovery settlement is then recorded as the load is reduced. Load increment is then increased and the procedure is repeated until the ground fails.

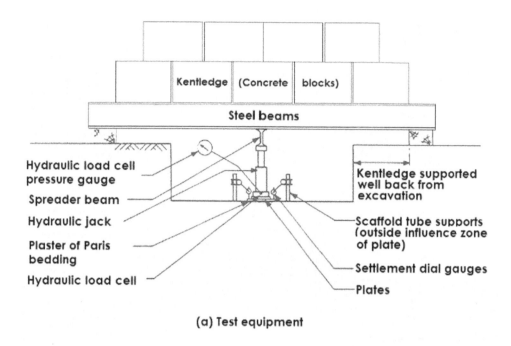

**(a) Test equipment**

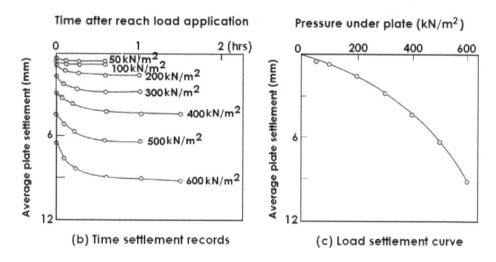

**(b) Time settlement records**   **(c) Load settlement curve**

*Figure 11.21* Plate loading test layout and result

   The plate test, however, is expensive and therefore is rarely used, unless information that is required could not be obtained by other means. This test is useful to predict settlement and bearing capacity of material such as soft weathered rock or road pavement and runways.
   Installation works for the plate has to be done with great care. All fine soils have to be removed from the excavated base. The plate is to be set in mortar or plaster of Paris. If this is not done, mostly likely the settlement measured is error that is not the actual deformation of the ground.

### 11.3.10 In situ permeability test

Although laboratory consolidation test is usually used to estimate magnitude of settlement, such data are not always reliable to predict rate of settlement especially for cases of soil with micro fabrics. Because of this the in situ permeability test is thought to be more suitable to determine the drainage characteristics of soil for predicting the rate of settlement. The techniques involve installation of piezometer or similar instruments in each layer of soil where the permeability value is required.

The test usually done is the falling head or constant head test. In general, the constant head test is easier to perform. This test is carried out with hydraulic piezometer or similar instrument. Rate of flow ($q$) into the soil through the piezometer tip is plotted as ordinate versus $1/t^{1/2}$ as abscissa ($t$ is time that elapse after start of test). In situ permeability can be obtained by extrapolating the initial straight-line part of the graph until it cut the ordinate axis where $q_\infty = k\,F\,\Delta h$ ($\Delta h$ is constant excess head, $F$ is shape factor of piezometer tip and $k$ is soil coefficient of permeability).

Most hydraulic piezometers are in the form of a cylinder of length $L$ and diameter $D$. Hvorslev (1951) gave the following equation for shape factor $F$:

$$F = \frac{2\pi L}{\ln\left[\dfrac{L}{D} + \sqrt{1 + \left(\dfrac{L}{D}\right)^2}\,\right]} \tag{11.7}$$

Coefficient of consolidation ($c_v$) can also be obtained from this test using the following equation:

$$c_v = q^2 r^2 / (pn^2) \tag{11.8}$$

where

  $r$ = radius of piezometer sphere
  $n$ = slope of $q - 1/t^{1/2}$ relation.

As alternative $c_v$ can be obtained by combining permeability, $k$ from field test with coefficient of volume compressibility ($m_v$) from laboratory test as follows:

$$c_v = k / (g_w m_v) \tag{11.9}$$

where $\gamma_w$ = unit weight of water.

### 11.3.11 Full-scale loading test

Study on trial embankment or structures of the same order as the proposed structure is called a full-scale loading test. This test is required in cases where the subsoil or structure is too complex for normal test and analysis.

However this test is expensive and is only carried out after considering normal methods as unsuitable.

The objectives of doing a full scale test are as follows:

1. To determine magnitude and duration of consolidation settlement and secondary compression for soil with micro fabric and anisotropy that do not allow representative tests to be performed in the laboratory.
2. To evaluate embankment stability for safe rate of loading or absolute shear strength of subsoil with loading to failure.
3. To study effectiveness of an improvement scheme, say a vertical drainage system.
4. To determine magnitude and rate of adjacent horizontal displacement.
5. To evaluate effectiveness of various construction procedures.

In general the most difficult site will be chosen. Choice is often added with data from thorough site investigation. The subsoil has to have similar composition with the rest of the project. Depending on the objective of the test, some or all of the observations below are required.

1. Settlement observation with settlement gauges.
2. Observation of vertical and horizontal deformation. Surface movement can be obtained by observing level of observation plate with ordinary surveying method. Horizontal deformation is usually monitored with an inclinometer.
3. Observation of pore water pressure with piezometer.

### 11.3.12 JKR probe

The JKR probe is widely used for site investigation works in addition to other more comprehensive site investigation work such as drilling. The probe was first introduced to the Public Works Department (JKR), Malaysia in 1972, based on principles outlined by Hvorslev (1949) for driven rod for sounding work.

The JKR probe is a simple site investigation procedure that is relatively inexpensive. It is often used in Malaysia, especially for preliminary site investigation, to assess the subsoil layer and the bearing capacity of soils. The probe comprises a $60°$ cone, 25 mm in diameter, screwed to the end of a rod. The rods are made of steel of type HY55C 12 mm in diameter, each 1.2 m in length and connected with a coupling of 22 mm external diameter. Driving is made using a small hammer weighing 5 kg falling from a height of 280 mm along a guide rod. The number of hammer blows to drive the cone through a distance of 300 mm is recorded (like the N value in the standard penetration test). This number of blows gives a measurement of soil consistency for cohesive soil and soil density for granular soil.

Figure 11.22 shows general arrangement of the JKR probe. Ooi and Ting (1975) have done research work with the JKR probe and produced a design chart for shallow footing as shown in Figure 11.23.

### 11.3.13 Geophysical method

This method is generally used for projects such as road construction. The main aim of the method is obtain information about soil and rock stratification over a large area quickly compared with the boreholes. However, the results obtained can be quite subjective and difficult

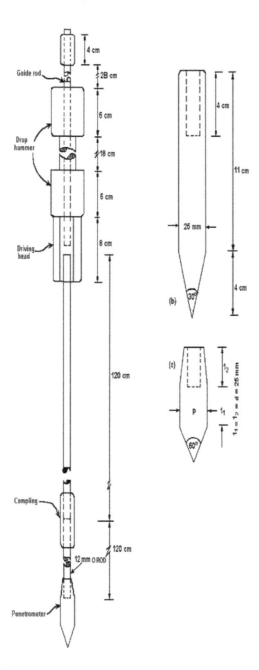

*Figure 11.22* JKR probe

to interpret clearly. Especially for case where a soft layer is present below a hard layer or for cases where the boundary between soil and rock could not be differentiated clearly.

There are two geophysical methods available: the seismic refraction method and electrical resistance. Of these two, the seismic refraction method is normally used.

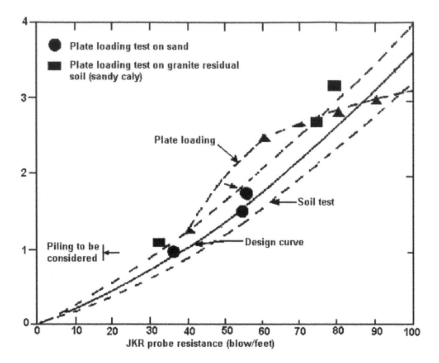

*Figure 11.23* Allowable bearing capacity versus JKR probe resistance

Source: Ooi and Ting (1975)

The seismic refraction method is based on the velocity seismic waves traveling through soil and rock. In general the denser the material, the faster is the seismic wave.

In conducting the test, seismic wave is generated at the ground surface by exploding a small explosive or dropping a heavy hammer on a steel plate. Sensors called geophones are placed at a known distance from this energy source. Time required for the generated wave to reach the sensor is recorded. With this, the depth of the soil strata and rock can be estimated. Figure 11.25 shows the procedures of the method.

Since the wave is faster through rock compared with soil, the reflected wave will reach the geophones located further from the source first compared with the direct wave, which is also known as surface wave. Time required for the wave to reach each of the geophone is plotted as shown in Figure 12.26. Velocity $V_1$ is for the geophone located closest to the energy source. With this, the gradient of time curve versus distance is inversely proportional to velocity, that is:

$$V_1 = \frac{L_2 - L_1}{t_2 - t_1} \tag{11.10}$$

where

$V_1$ = velocity of wave through upper soil layer
$L_1$ and $L_2$ = distance of geophones 1 and 2 from energy source
$t_1$ and $t_2$ = time required for the wave to reach geophones 1 and 2.

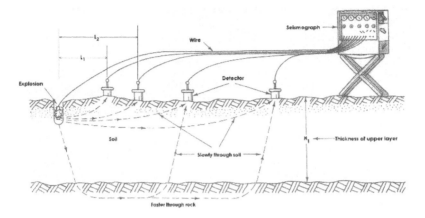

*Figure 11.24* Seismic refraction method

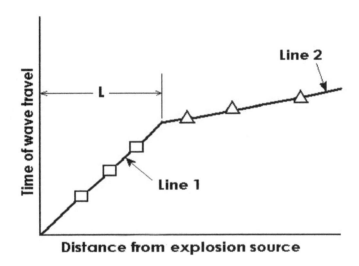

*Figure 11.25* Plot of seismic refraction test

So is the case with $V_2$, which is the gradient of line 2 in Figure 11.24. Thereby thickness of stratum $H_1$ can be estimated from the following equation:

$$H_1 = \frac{L(V_2 - V_1)^{\frac{1}{2}}}{2(V_2 + V_1)}$$

(11.12)

where

$H_1$ = thickness of upper layer
$L$ = distance of intersection point of two slopes plotted (see Figure 11.25).

A list of velocity for various types of soil and rock is given in Table 11.6.

*Table 11.6* Representative values of velocity (feet/second)

| | |
|---|---|
| *Unconsolidated materials* | |
| Most unconsolidated materials | below 3000 |
| Soil | |
| normal | 800–1500 |
| hard | 1500–2000 |
| Water | 5000 |
| Loose sand | |
| Above groundwater level | 800–2000 |
| Below water level | 1500–4000 |
| Mixture of sand and loose gravel, wet | 1500–3500 |
| Loose gravel, wet | 1500–3000 |
| *Consolidated material* | |
| Most hard rock | above 800 |
| Coal | 3000–5000 |
| Clay | 3000–6000 |
| Shale | |
| Soft | 4000–7000 |
| Hard | 6000–10000 |
| Sandstone | |
| Soft | 5000–7000 |
| Hard | 6000–10000 |
| Limestone | |
| Weathered | below 400 |
| Hard | 8000–18000 |
| Basalt | 8000–13000 |
| Granite and gneiss | 10000–20000 |
| Frozen soil | 4000–700 |

## 11.4   Field instrumentation

Field instrumentation is used to obtain measurements of certain soil parameters such as pore water pressure, stresses in soil, strain and deformation that occur when the soil is loaded. These instrumentations, however, are mostly used for research and large projects only.

Besides for measurement of soil parameters, field instrumentation is also used to check the assumptions made in designing a particular structure. By measuring behavior of the actual structure, comparison between the actual behaviors with the design prediction could be made.

As stated above, soil properties that are normally measured by field instrumentation are pore water pressure, soil stress (either total or effective stress), soil deformation, either vertical or horizontal deformation, measurement of load and strain components of structure.

There are several instrumentation systems that are available in the market. Most of these instruments, however, are more or less using the same principle. Each has its advantages and limitations. In this chapter we will briefly describe the instruments used to measure parameters such as pore water pressure, soil stresses, deformation, strain, and load.

### 11.4.1   Pore pressures

There are two general categories of equipment used for measuring pore water pressure. They are the well or observation well for measuring groundwater level, and piezometer for measuring pore water pressure. For the case of the piezometer, there are five types that are normally being used.

1.   Open standpipe
2.   Double tube hydraulic piezometer
3.   Pneumatic piezometer
4.   Vibrating wire gauge piezometer
5.   Fixed resistance strain gauge piezometer.

### 11.4.2   Observation well

A cross-section of an observation well is shown in Figure 11.26. The well comprises a perforated section connected to another pipe. This pipe is inserted into a borehole, which is filled with gravel or sand. The groundwater level is measured with a sounding tape.

### 11.4.3   Open standpipe piezometer

The shape of this piezometer is almost similar to that of an observation well. The only difference is that the perforated end is sealed so that the pore pressure at depth of measurement (localised) will not be influenced by the pore pressure at other levels. As shown in Figure 11.27, the perforated end is connected to a standpipe.

Water level in the standpipe represents pore pressure head at which the perforated end is placed. A tape or sounding probe measures this head.

In soft ground, a standpipe piezometer can be installed by driving a perforated end attached to a pipe or drilling rod. These pipes can be withdrawn if need be. But this method needs to be used with great care as it does not have special sealing materials. The seal is provided by the soft soil only. Adequate sealing is required in order for the piezometer to function properly.

### 11.4.4   Double tube hydraulic piezometer

Figure 11.28 shows a double tube hydraulic piezometer system, which is also sometimes known as closed hydraulic piezometer. The system comprises a perforated end bit connected to two standpipes or plastic tube, one of which is attached to a pressure meter. Meanwhile, the second tube is used to discharge any gas or air that might enter into the piezometer system. These tubes can be placed horizontally or connected to outside the construction area. To obtain accurate readings, air bubbles must be removed from the tube. Water used for cleaning the system must not contain air.

### 11.4.5   Pneumatic piezometer

Figure 11.29 shows a pneumatic piezometer system, which comprises a sensor connected to two tubes one of which (in flow channel) is connected to a pressure gauge and a gas supply,

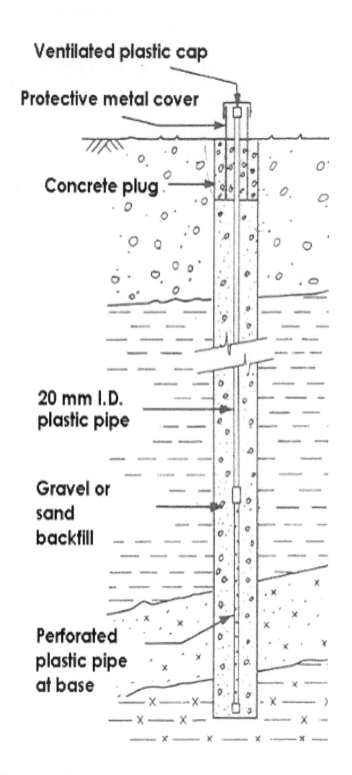

Figure 11.26    Observation well

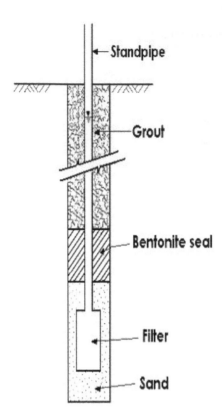

Figure 11.27 Open standpipe piezometer

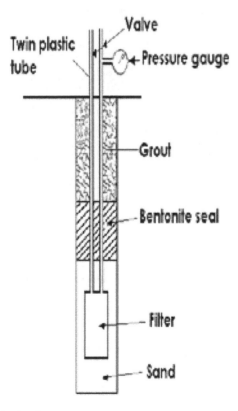

Figure 11.28 Double tube hydraulic piezometer

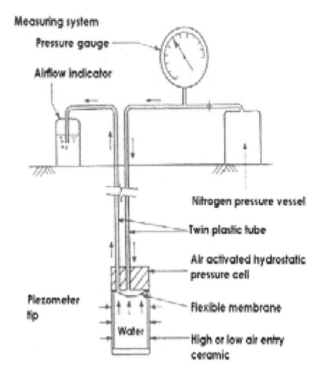

*Figure 11.29* Pneumatic piezometer system

Source: Clayton et al. (1982)

while the second tube (outflow channel) is attached to leveling equipment and exposed to the atmosphere.

The sensor consists of a control valve which controls the inflow and outflow, flexible membrane which separates the water from the measuring system, and porous stone which allows water to flow to the membrane.

The sensor attached to a pipe is inserted into a borehole to the level required. The hole is then sealed with bentonite and grout.

Supplying compressed air into the inflow channel allows pore water pressure reading. When this pressure is in equilibrium with the pore pressure, the control valve will open to allow gas out through the outflow channel. A pressure gauge inside the inflow channel will record the pressure of gas in it. Either air or nitrogen is normally used. A number of piezometers use oil instead of gas, but this type of piezometer does not show any advantage compared with the gas piezometer.

### 11.4.6   Vibrating wire strain gauge piezometer

Vibrating wire strain gauge piezometer has a metal membrane that forms a separator between water and the sensor system. The sensor system comprises a pretension wire placed at the center of a membrane as shown in Figure 11.30. With this any change in the pore water

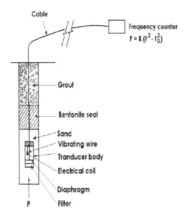

## (a) Schematic of a vibrating wire piezometer

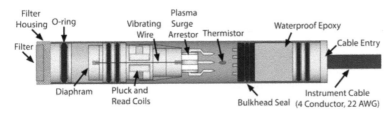

(b)  Section of a vibrating wire piezometer (Model 4500S. Courtesy of Geokon, USA)

*Figure 11.30* Vibrating wire piezometer

pressure will deform the membrane and change the tension of the wire. Plucking the wire with an electric vibrator and measuring its frequency measures the tension of the wire.

Wire will vibrate in a magnetic field and induce alternating voltage flow across the membrane. The frequency of the voltage flow is similar to the frequency of the vibrating wire and can be measured with the frequency sensor. Pore water pressure is then read from the frequency manual or calibrated curve.

### 11.4.7   Bonded electrical resistance piezometer

This bonded electrical resistance or fixed strain gauge of piezometer (see Figure 11.31) has a transducer, which is in direct contact with the pore water. Fixed resistance strain gauge is placed on the membrane. Any strain induced in the membrane will be measured directly by the strain gauge. With this, the signal output from the transducer can be used to directly measure the pore pressure.

### 11.4.8   Hydrostatic lag time

For a piezometer to accurately record pore pressure change, there should be no inflow or outflow of water from the piezometer. However, when there is change in pressure, pore water

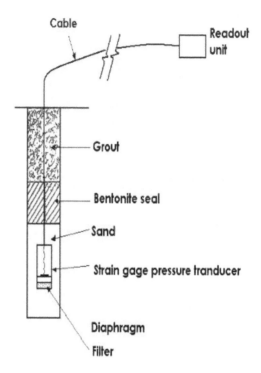

*Figure 11.31* Fixed resistance strain gauge piezometer

will flow to the piezometer for case of pressure increase, and from the piezometer in case of reduced pressure. A certain amount of time will therefore have to lapse to achieve a state of equilibrium (no more flow) after a particular flow. This time is known as the hydrostatic lag time. Main factors, which influence the hydrostatic lag, are soil compressibility, type and size of piezometer and magnitude of pore pressure change. Standpipe piezometer, for instance, has a bigger lag time compared with a vibrating wire piezometer because of the larger water flow that has to occur in the standpipe piezometer, compared with the vibrating wire piezometer. Methods to calculate this hydrostatic lag time have been proposed by Terzaghi and Peck (1967) and Hanna (1973).

Advantages and disadvantages of each of the instruments used for measuring pore water pressure are summarised in Table 11.7.

### 11.4.9   Earth pressure

Measurement of earth pressure can be divided into two main categories: (a) measurement of total stress at any point in the soil mass and (b) measurement of total and contact stress between soil and structure. The sensor system of earth pressure cell, in general, is rather like the sensor of a vibrating wire piezometer.

There are several types of earth pressure cells: pneumatic, hydraulic, vibrating wire strain gauge, fixed resistance strain gauge and unrestrained resistance strain gauge.

*Table 11.7* Instruments for measuring groundwater pressure

| Instrument Type | Advantages | Limitation |
|---|---|---|
| Observation well | Easy to install and low cost. | |
| Open standpipe piezometer | Easy to install, low cost and reliable. Long successful performance record. Self-deairing if inside diameter of standpipe is adequate (>10 mm). | Porous filter can be plug. Long time lag. Subject to damage by construction equipment and by vertical compression of soil around standpipe. |
| Twin-tube hydraulic piezometer | Easy to use and reliable. Long successful performance record. Can be used to measure permeability. | Periodic flushing may be required. Tubing must not be significantly above minimum piezometric elevation. |
| Pneumatic piezometer | Stable and short time lag. Minimum interference to construction. Calibrated part of system accessible. | Expensive Using dry gas. Several models needed log time lag. |
| Vibrating wire piezometer | Easy to read. Short time lag. Can be used to read negative pore water pressure. | Not suitable for dynamic reading. Need for lightning protection should be evaluated. |
| Bonded electrical resistance piezometer | Easy to read. Short time lag. Minimum interference to construction. Suitable for dynamic measurements. Can be used to read negative pore pressures. | Long-term stability uncertain. Low electrical output. Lead wire effects. Errors caused by moisture, temperature and electrical connections are possible. |

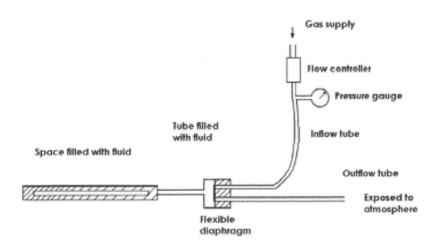

*Figure 11.32* Pneumatic earth pressure cell

In general earth pressure cell comprises a chamber under pressure. The pressure is transferred to the fluid inside the chamber, which is then measured by a sensor system. Figure 11.32 shows a pneumatic earth pressure cell.

Measurement of the total stress of a particular soil mass requires the following conditions to be satisfied:

1.  The presence of the pressure cell will not grossly modify the stress condition. This disturbance, however, could not be totally avoided.
2.  Size and surface of the sensor has to be sufficient to reduce effect of local non-homogeneity, and cell will record average stress of the area.
3.  The method of installation does not significantly modify the soil state of stress. This third condition limits the use of earth pressure cell for earth fill only. However, some success has also been obtained in measuring horizontal stresses in soft clays.

Measurement of stress imposed by a particular structure requires the pressure cell to act like the structure itself. With this the effect of arching over the cell should be minimised. Arching is due to the different deflection of both cell and structure. Another problem is that contact surface between the cell and the structure which may not be uniform, thereby resulting in variation and non-permanency in stresses recorded. This sort of data is difficult to interpret.

There are many factors that influence the reading of a pressure cell. Results of study by several researchers show that the cell should:

1.  Be as thin as possible in size compared with its surface area in direction of stresses to be measured
2.  Be more robust than the soil
3.  Have measuring surface many times larger than the soil particle size
4.  Be calibrated for effect of rotation major principal stress
5.  Have simple geometry
6.  Be durable so that soil around it can be compacted during installation
7.  Not easily rust and damage.

In addition to the above, installation of the cell also affects the earth stress reading.

Normally the cell is placed in a hole that is then compacted by hand or light machine, whereas the fill is compacted with heavy machineries. In any case, this method is used in order to prevent damage to the gauge. It is also the best practical method currently available, although the stresses reading recorded might be lower than the actual stress.

### 11.4.10    Measurements of deformation

There are several methods and instruments that could be used for measuring horizontal, vertical, and axial deformation. The types of equipment and their categories are summarised in Table 11.8.

### 11.4.11    Surveying method

Surveying methods normally used include optical leveling, offset from transit line, triangulation, photogrammetry and electronic distance measurements (EDM). All settlement and heave measurement must be made with reference to a benchmark, while horizontal movement is referred to a stable reference point.

*Table 11.8* Categories of instruments for measuring deformation

| Category | Type of measured deformation | | | | |
|---|---|---|---|---|---|
| | ↔ | ↕ | ↗ | ⟂ | ⊤̄ |
| Surveying methods | • | • | | • | |
| Portable deformation gauge | | | • | • | |
| Settlement markers | • | • | | • | |
| Standpipe settlement gauge | | • | | | • |
| Heave gauge | | • | | | • |
| Inclinometer | | • | | | • |
| Borehole strain meter | • | | • | | • |
| Soil strain gauge | | | • | | • |

↔ Horizontal deformation    ⟂ Surface

↕ Vertical    ⊤̄ Subsurface

↗ Axial

*Table 11.9* Advantages and limitations of the deformation gauge

| Instrument | Accuracy | Advantages | Limitations |
|---|---|---|---|
| Graduated scale | ± 0.5 mm | Easy to use and inexpensive | Resistance and accuracy are limited |
| Measuring tape | ± 2.5 mm | Easy to use and inexpensive | Accuracy limited |
| Strain gauge Mechanical | ± 0.005 mm −0.013 mm | Precise | Very short span |
| Micrometer or dial gauge | ± 0.0002 mm | Easy and inexpensive | Short span |
| Portable tape or elongation meter rod | ± 0.025 mm | Easy and precise | Accuracy limited caused by flexibility |

## 11.4.12 Portable deformation gauge

A portable deformation gauge is normally used to monitor changes in cracks in buildings or ground surface. The advantages and limitations of the deformation gauge are summarised in Table 11.9.

## 11.4.13 Settlement markers

Settlement markers are used for monitoring ground surface deformation, in conjunction with the surveying method. The markers are installed on the original ground surface or on a structure. Two types of markers are normally used. They are the settlement rod and control rod. The settlement rod is made of steel, which is inserted into a pipe casing.

Control rod comprises a T-shaped piece of wood which is inserted into the ground. The problem however is that this rod is easily disturbed by the construction activity and therefore has lack of accuracy.

### 11.4.14   Standpipe settlement gauge

Standpipe settlement gauge is used to measure settlement of subsoil layer, fill or surcharge. The gauge is in the form of a pipe or rod with one of its ends placed on the soil layer whose settlement is to be measured. The other is upright up to the soil surface. The level of the rod head indicates settlement that has occurred.

Table 11.10 shows several types of standpipe gauges together with their advantages and limitations. These gauges are shown in Figures 11.33–11.36.

Table 11.10 Several types of standpipe gauges together with their advantages and limitations

| Type | Accuracy | Advantages | Limitation | Reliability |
|---|---|---|---|---|
| Settlement platform | ±2.5–25 mm | Simple | One point observation only. Contain cumulative errors due to long pipe. | Very good |
| Borros anchor | ±2.5–25 mm | Simple | One point observation only. Contain cumulative errors due to long pipe. Cannot be used in soft oil. | Good |
| Spiral-foot anchor | ±2.5–25 mm | | Can be used with Borros anchor. | Very good |
| Inductively loop | ±2.5–25 mm | Can be used in multipoint borehole | Readings are subjective. Accuracy depends on level of electric current. | Good |
| Telescopic settlement gauge | ±2.5–25 mm | Simple; multipoint observation | Not suitable to borehole. | Good |

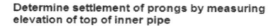

**Determine settlement of prongs by measuring elevation of top of inner pipe**

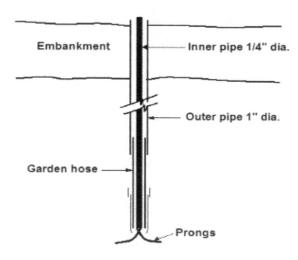

Figure 11.33  Borros anchor

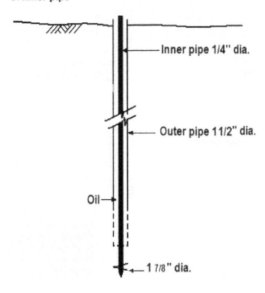

Determine settlement of foot by measuring elevation of top of inner pipe

Inner pipe 1/4" dia.

Outer pipe 1 1/2" dia.

Oil

1 7/8" dia.

Figure 11.34 Spiral-foot anchor

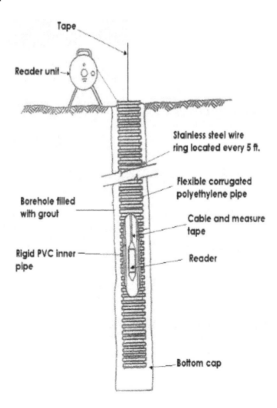

Tape

Reader unit

Stainless steel wire ring located every 5 ft.

Flexible corrugated polyethylene pipe

Borehole filled with grout

Cable and measure tape

Rigid PVC inner pipe

Reader

Bottom cap

Figure 11.35 Inductively loop

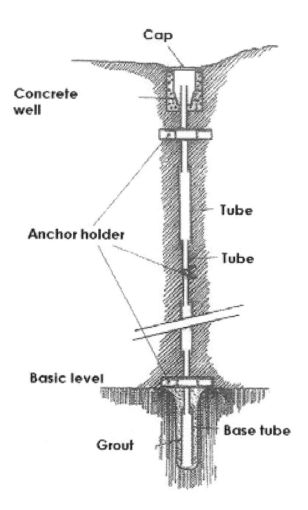

*Figure 11.36* Telescopic settlement gauge

### 11.4.15  Heave gauge

A heave gauge is used to monitor heave that occurs at the soil base. A simple technique is illustrated in Figure 11.37. A pipe with a conical shape head is placed at base of borehole, with its head pointing upwards. The level of the excavation base is determined by leveling the cone head. The accuracy of this technique is between ±5–25 mm. Reading should no longer be made when borehole side has collapsed.

### 11.4.16  Liquid level system

Settlement beneath embankment, fills or heave can be measured by monitoring any change in elevation between a reservoir and settlement plate as shown in Figure 11.38. The liquid reservoir is ideally mounted on a stable location. As the plate settles, liquid pressure at the settlement plate increases and is monitored by a vibrating wire sensor.

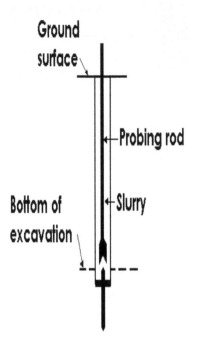

Figure 11.37 Heave gauge

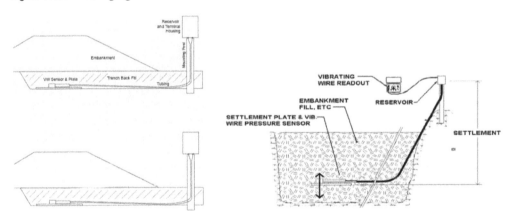

(a) Schematic of a liquid level settlement system

(b) Image of the system reservoir
(photo courtesy of RST Instruments Ltd., Canada)

Figure 11.38 Liquid level settlement system

### 11.4.17   Inclinometer

Inclinometer is normally used to measure horizontal movements below ground surface. An inclinometer system comprises a pipe with guide channel and an electric sensor, as shown in Figure 11.39.

Electric probe sensor is lowered into the tube with its rod guided by the guiding rod. Sensor attached to a reading unit will record the inclination and depth of the sensor enabling lateral deformation of the ground to be calculated.

### 11.4.18   Soil strain gauge

A soil strain gauge is usually used to measure strain in earth fill. This instrument uses the principle of inductance between two coils buried inside the ground. The two coils are related to one another with their positions as shown in Figure 11.40. One of the coils is attached to electromagnetic inducers resulting in electromagnetic field around it. The second coil will be induced by the electromagnetic field to yield electric current of magnitude inversely proportional to the distance between the two coils. Soil strain is determined by measuring changes in distance between the two coils, after the installation was made.

The main advantage of this method is that its installation is simple because there is no mechanical connection between the two coils. Dynamic strain may also be measured with this instrument. The obvious limitation however is that the instrument is sensitive to the presence of metals in its surroundings.

### 11.4.19   Fiber-optic sensors

Fiber-optic sensors are optical fibers that work either as a sensing element or as a means of relaying signals from a remote sensor to the electronics that process the signals. Depending on the application, fiber may be used because of its small size, or because no electrical power is needed at the remote location. The sensors are used to measure strain, temperature, pressure and other quantities by modifying a fiber so that the quantity to be measured modulates the intensity, phase, polarisation, wavelength or transit time of light in the fiber. Meanwhile a distributed fiber-optic sensing is a technology that enables continuous real-time measurements along the entire length of a fiber-optic cable. These sensors have been developed for a range of civil engineering monitoring applications.

## 11.5   Choice of instruments

In the preceding section we explained the various instruments that could be used to measure various soil parameters. To measure the pore water pressure, for example, various types of piezometers are available that could be used. A choice of the most suitable instrument therefore often has to be made. Among the main considerations that must be accounted for when choosing the instruments are:

1.   Reliability of the instruments, in that they cannot be easily damaged
2.   Accuracy
3.   Cost
4.   Purpose of installing the instrument.

Figure 11.41 shows a typical instrumentation scheme of an embankment

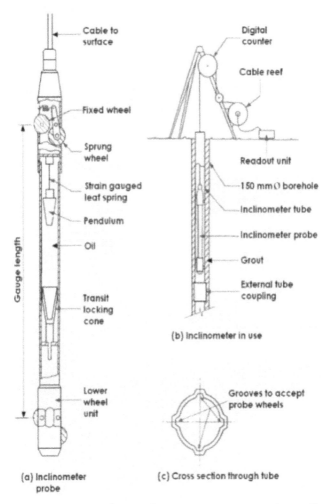

Cable to surface

Digital counter

Cable reef

Fixed wheel

Sprung wheel

Readout unit

Strain gauged leaf spring

150 mm Ø borehole

Inclinometer tube

Pendulum

Inclinometer probe

Oil

Grout

Transit locking cone

External tube coupling

(b) Inclinometer in use

Lower wheel unit

Grooves to accept probe wheels

Gauge length

(a) Inclinometer probe

(c) Cross section through tube

(a) Schematic of an inclinometer section and installation

(b) Image of an inclinometer equipment
(Courtesy of RST Instruments Ltd., Canada)

*Figure 11.39* Inclinometer equipment
Source: Soil Instruments Ltd

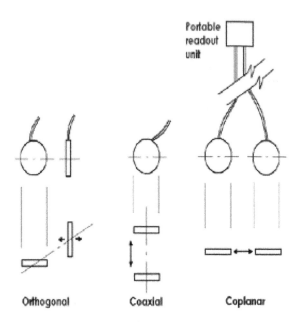

Figure 11.40 Soil strain gauge configuration

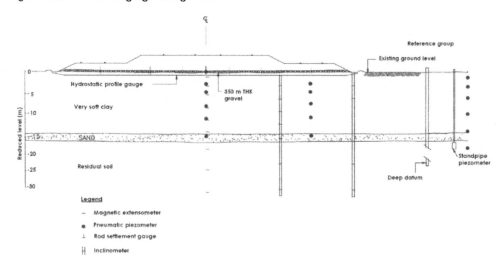

Figure 11.41 Typical instrumentation scheme of an embankment

## 11.6  Assessment and control

As mentioned earlier, field instruments can be used not only for measuring soil parameters but also to check the assumptions made in design of a particular structure. By measuring the behavior of the actual structure, comparison between the actual behaviors with the design prediction could be made. Field instruments can also be used to control rate of construction, such as to monitor generation and dissipation of excess pore water pressure, in order to ensure safe construction.

Perhaps the simplest and most widely used field instruments are those for measuring the current and then predicting future settlements of the ground.

The technique of predicting future settlements based on the current observed values is known as the observational procedure. Usually two methods are used. They are the Asaoka method and the hyperbolic method.

Asaoka (1978) observational procedures use early field settlement data to predict end-of-primary settlement and in situ coefficient of consolidation. This method has been enjoying increasing popularity especially for consolidation of clays. In the Asaoka method the one dimensional consolidation settlement $S_1$, $S_2$, $S_3$, $S_{n-1}$, $S_n$ and so forth at times 0, $dt$, $2dt$ and so forth (i.e., at equal time increments) can be expressed, as a first order approximation, by:

$$S = a + \beta S_{n-1} \tag{11.13}$$

Equation (11.13) represents a straight line when values of $S_n$ are plotted on the vertical axis and values of $S_{n-1}$ are plotted on the horizontal axis. Hence $\alpha$ represents the intercept on the vertical axis and $\beta$ the gradient.

A typical straight-line plot that emerges when value of $S_{n-1}$ are plotted against $S_n$ for a series of equal time intervals is illustrated in Figure 11.42.

From the settlement-time curve in Figure 11.42, when settlement is complete, $S_n = S_{n-1}$. The equilibrium line $S_n = S_{n-1}$ is the straight line drawn at 45° (i.e., $b = 1$).

The ultimate (100%) settlement, $S_{100}$, can be obtained by substituting $S_n = S_{n-1} = S_{100}$ into Equation (11.10),

$$S_{100} = a + \beta S_{100}$$

and

$$S_{100} = \alpha / (1 - \beta) \tag{11.14}$$

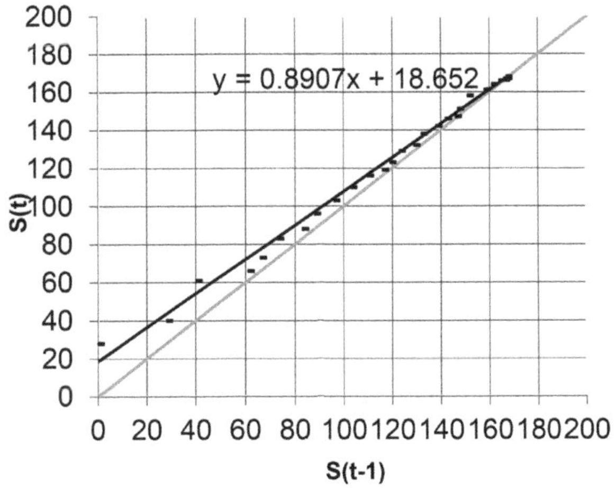

Figure 11.42 Typical Asaoka plot

The 90% settlement ($S_{90}$) is thus given by:

$$S_{90} = 0.9\alpha / (1 - \beta) \qquad (11.15)$$

The number of time increments ($J_{90}$) needed to achieve 90% settlement is given by:

$$J_{90} = In(1 - U_{90}) / In\beta \qquad (11.16)$$

The foregoing provides the basis for utilising settlement data to make assessments of the degree of settlement that is occurring within the monitored area. Referring to the straight-line Asaoka plot shown in Figure 11.42, the value $S_{100}$ is obtained when the best fit straight-line through the site data allows increasingly refined predictions of the magnitude and rate of total settlement to be made.

For the case of clay soils, this method seems to work rather well (Huat, 2002). For peat soils, Cartier et al. (1989) used Asoka's method for the analysis of a test embankment on peat and reported a reasonable prediction of settlement and time of 98% primary consolidation. Edil et al. (1991), on the other hand, applied the procedure to a variety of clay and peat cases and questioned its applicability to peat settlement.

Another useful method is the hyperbolic method (Tan, 1971; Chin, 1975). This method is based on the assumption that the settlement-time curve is similar to a hyperbolic curve and can be represented by the equation (see also Figure 11.43).

$$S = \frac{t}{(c + mt)} \qquad (11.17)$$

where $S$ is the total settlement at any time after the excess pore water pressure has dissipated and $m$ and $c$ are empirical constants. The plot with the ratio of $t/S$ on the ordinate and time $t$ on the abscissa, a straight line gives intercept $c$ with slope $m$, and the significance of $m$ by writing the equation is as follows:

$$\frac{1}{S} = m + \frac{c}{t} \qquad (11.18)$$

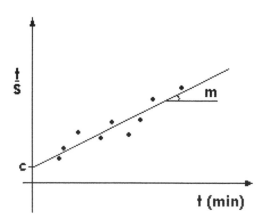

*Figure 11.43* Hyperbolic method to predict future settlement

*Table 11.11* Comparison of predicted and measured settlements of Tangkak (Malaysia) trial embankment (Huat, 2002)

| $t_0$ | ¼ | ½ | 1 | 2 | 3 | 4 | 5 | 6 | 7 | 8 | 10 | 20 | 30 |
|---|---|---|---|---|---|---|---|---|---|---|---|---|---|
| *r* | | | | | | | | | | | | | |
| ¼ | | 744 | 918 | 1041 | 1089 | 1115 | 1131 | 1142 | 1150 | 1156 | 1165 | 1182 | 1188 |
| ½ | | | 820 | 901 | 932 | 948 | 958 | 965 | 969 | 973 | 978 | 986 | 992 |
| 1 | | | | 1092 | 1185 | 1238 | 1272 | 1296 | 1313 | 1326 | 1346 | 1386 | 1400 |
| 2 | | | | | 1089 | 1158 | 1204 | 1236 | 1261 | 1379 | 1306 | 1365 | 1385 |
| 3 | | | | | | 1192 | 1233 | 1262 | 1283 | 1299 | 1324 | 1374 | 1392 |
| 4 | | | | | | | 1267 | 1306 | 1335 | 1357 | 1389 | 1460 | 1485 |
| 5 | | | | | | | | 1228 | 1259 | 1283 | 1318 | 1395 | 1423 |
| 6 | | | | | | | | | 1350 | 1386 | 1434 | 1542 | 1581 |
| Actual settlement (mm) | 550 | 670 | 835 | 1010 | 1150 | 1230 | 1300 | 1350 | 1390 | 1400 | | | |

Indicates a prediction within 10% of measured value

$t_0$ = time after start of construction (in year)
$\rho$ = predicted settlement in mm

When *t* becomes very large (i.e., toward infinity), $1/t \to 0$ and then $1/S = m$, which means the ultimate settlement, $S_{ult} = 1/m$.

The hyperbolic method is apparently good for predicting future settlements especially if start of construction data is used, and after more than 50% of the settlement has occurred (Huat, 2002). An example is shown in Table 11.11. In the case of peat, studies carried out by Al-Raziqi et al. (2003) showed that this method could be used for predicting the primary phase of peat settlement. However for the secondary (creep) phase, the settlement prediction could be misleading.

## References

Al-Raziqi, A. A., Huat, B.B.K. & Munzir, H. A. (2003) Potential usage of hyperbolic method for prediction of organic soil settlement. In: Huat et al. (eds) *Proceedings of 2nd international conference on advances in soft soil engineering and technology*. Universiti Putra Malaysia Press, Putrajaya, Malaysia. pp. 439–445.

Asaoka, A. (1978) Observational procedure of settlement prediction. *Soils and foundation*, 18(4), 87–101.

Begemann, H. K. S. (1965) The friction jacket cone as an aid in determining the soil profile. *Proceedings of the 6th International Conference on Soil Mechanics and Foundation Engineering*. University of Toronto Press, Montreal, 1, pp. 17–20.

Bishop, A. W. (1948) A new sampling tool for use in cohesionless sands below ground water level. *Geotechnique*, 1, 125–131.

Bjerrum, L. (1973) Problem of soil mechanics and construction on soft clays and structurally unstable soils. State-of-the-art report, Session IV. *Proceedings of the 8th International Conference Soil Mechanics and Foundation Engineering*. Moscow, Vol. 3, pp. 111–159.

BS 5930. (2015) Code of Practice for Ground Investigations, British Standards Institution.

Cartier, G., Allaeys, A., Londez, M. & Ropers, F. (1989) Secondary settlement of peat during a load test. *Proceedings of 12th International Conference on Soil Mechanics and Foundation Engineering*. Rio de Janeiro, Brazil, CRC Press/Balkema, Rotterdam, Netherlands, 3, pp. 1721–1722.

Chin, F. K. (1975) The seepage theory of primary and secondary consolidation. *4th Southeast Asian conference on soil engineering*. Kuala Lumpur. pp. 21–28.

Clayton, C.R.I., Simons, N. E. & Matthews, M. C. (1982) *Site investigation*. Granada Publishing, St Albans, Hertfordshire, England.

Edil, T. B., Fox, P. J. & Lan, L. T. (1991) End-of-primary consolidation of peat. *Proceedings 10th ECSMFE*. A.A. Balkema, Florence, 1, pp. 65–68.

Ehrenberg, J. (1933) Gerätezurentbahme von boden proben für boden physikalischer Unter suchungen. *Bautechnik*, 11, 303–306.

Gibson, R. E. & Anderson, W. F. (1961) In situ measurement of soil properties with the pressure meter. *Civil Engineering and Public Works Review*, 56(658), 615–618.

Hanna, T. H. (1973) *Foundation Instrumentation*. Trans Tech Publication, Cleveland, USA.

Huat, B.B.K. (2002) Hyperbolic method for predicting embankment settlement. *2nd World Engineering Congress*. Universiti Putra Malaysia Press, Kuching, Sarawak, 1, pp. 228–232.

Huizinga, T. K. (1944) Tienjaregrondmechanica in Nederland. *Weg en Waterbouw*, 1–2 August.

Hvorslev, M. J. (1949) Preliminary draft report on the present status-of-the-art of obtaining undisturbed samples of soils. *Supplement to Proceedings of the Purdue Conference on Soil Mechanics and Its Applications*. Purdue University Press, Lafayette.

Hvorslev, M. J. (1951) *Time Lag and Soil Permeability in Ground Water Observations*. Bull No. 36. Waterways Experimental Station, Vicksburg, MS, USA.

Ooi, T. A. & Ting, W. H. (1975) The use of light dynamic cone penetrometer in Malaysia. *Proceedings of the 4th South East Asian Conference Soil Engineering*. Kuala Lumpur, Malaysia.

Peck, R. B., Hanson, W. E. & Thornburn, T. H. (1974) *Foundation engineering*. John Wiley, New York.

Robertson, P. K. (2010) Soil behaviour type from the CPT: An update. *2nd International Symposium on Cone Penetration Testing*. CPT'10, Huntington Beach, CA, USA. Available from: www.cpt10.com

Robertson, P. K., Campanella, R. G., Gillespie, D. & Greig, J. (1986) Use of piezometer cone data. *In-Situ '86 Use of In-situ testing in Geotechnical Engineering*. GSP 6, ASCE, Reston, VA, Specialty Publication, SM 92. pp. 1263–1280.

Robertson, P. K. & Capal, K. L. (2015) *Guide to Cone Penetration Testing for Geotechnical Engineering* (6th ed.). Gregg Drilling and Testing, Inc. Signal Hill, CA, USA.

Rowe, P. W. (1972) The relevance of soil fabric to site investigation practice. *12th Rankine Lecture, Geotechnique*, 22(2), 195–300.

Seaby, D. A. (2000) Portable, two stage sampler for difficult soils. *Soil Science Society of America Journal*, 64, 1327–1329.

Tan, S. B. (1971) Empirical method for estimating secondary and total settlement. *Proceedings of 4th Asian Regional Conference on Soil Mechanics and Foundation Engineering*. Asian Institute of Technology, Bangkok, Thailand, 2, pp. 147–151.

Terzaghi, K. & Peck, R. B. (1967) *Soil Mechanics in Engineering Practise*. John Wiley, New York.

# Index